Table of Contents

Table des Matieres

Opening address

Allocution d'ouverture

Ladies and Gentlemen,

As Director General of the OECD Nuclear Energy Agency I should like to welcome you today to this Seminar on the Bituminization of Low and Medium Level Radioactive Waste which NEA has organised in co-operation with Eurochemic. As most of you know, the Eurochemic Company is one of the joint undertakings of the OECD Nuclear Energy Agency, and was founded in 1959 by 13 NEA Member countries* with the purpose of designing, constructing and operating both a demonstration reprocessing plant and a research laboratory specialised in aqueous reprocessing of irradiated nuclear fuels and associated fields.

The reprocessing plant started hot operation in July 1966 and continued to operate until July 1974. During this period Eurochemic processed practically all types of available nuclear fuels and, as a result, has built up an almost unique know-how in the various aspects of fuel reprocessing, and also in waste management techniques. In fact, the major job ahead now for Eurochemic is the treatment of all the wastes resulting from these eight years of plant operation.

I am not going to expand on the subject of incorporation of radioactive waste into bitumen, which will be substantially discussed during the Seminar, and I would rather make some comment concerning radioactive waste management in general and the potential available at Eurochemic for research and development in this field. The present reprocessing situation, both in Europe and the USA, seems more complex than it appeared a few years ago, due particularly to the high burn-up fuels now used in light water reactors. As a consequence, the whole back-end of the fuel cycle, including waste management, has received increased attention and waste management has become a key factor in the licensing of reprocessing plants. Also, of course, the public has, rightly or wrongly, focussed their attention on waste management as one of the major problems in connection with the acceptance of nuclear energy.

In this general context, consultations are presently going on in order to set up an international research and development programme on the incorporation of high level waste into metal matrices, based on preliminary studies started a few years ago at Eurochemic. This programme has attracted significant interest from Eurochemic countries, and also from other countries such as Japan and the United States. We hope to be in a position to conclude a formal agreement between all interested countries in the next few months.

Eurochemic has also accumulated useful experience in the field of decontamination of the reprocessing cells which is of

---

* Austria, Belgium, Denmark, France, the Federal Republic of Germany, Italy, the Netherlands, Norway, Portugal, Spain, Sweden, Switzerland and Turkey.

primary importance not only for the possible later dismantling of the plant but also for the repairs and modifications associated with the normal operation of any reprocessing plant.

The management of solid wastes, particularly those contaminated by plutonium and other actinides, will also receive special attention in the near future and, again, this aspect of the radioactive waste problem is of special interest for the development of nuclear energy, notably in view of the possibility of plutonium recycling in present and future reactors. The treatment of organic waste such as spent solvents and the solidification of high level wastes at low temperatures through the formation of aluminium phosphates (known at Eurochemic as the LOTES process) are other examples of R and D to which Eurochemic can contribute usefully.

These few examples show how important is the know-how and experience gained by Eurochemic in all these aspects of waste management and the contribution that Eurochemic is in a position to offer in the general discussion of these problems. The bitumen facility at Eurochemic is only one of the aspects of this multifacet experience, and I hope there will be other opportunities in the future to draw attention to the other aspects.

However, the subject of this Seminar is of course bituminization and I should now like to declare the Seminar open and to wish you every success for your discussions and visits.

Session I

Chairman - Président

E. DETILLEUX

Séance I

PASSE, PRESENT ET FUTUR DU CONDITIONNEMENT DANS LE BITUME DES DECHETS RADIOACTIFS
HISTORIQUE DU PROCEDE

P. Dejonghe.
C.E.N./S.C.K.
Mol (Belgique).

INTRODUCTION

L'idée d'appliquer le bitume au conditionnement des déchets faiblement radioactifs date de juste avant la Conférence de Genève de 1958. Au départ, et encore maintenant, personne n'a cru que le bitumage était appelé à résoudre tous nos problèmes en matière de conditionnement et de stockage ou rejet. Au contraire, la technique était considérée comme une approche parmi d'autres et on en reconnaissait pleinement les limites d'application.
Néanmoins, l'idée a fait un certain chemin depuis presque 20 ans, que la plupart d'entre nous connaissent.
Nous aurons d'ailleurs le plaisir d'en apprendre davantage aujourd'hui et demain.

CONTEXTE INITIAL

Je crois que les premières applications du bitume comme moyen de conditionnement de déchets radioactifs furent étudiées au C.E.N./S.C.K. à Mol. Le motif de ces études était la recherche d'une technique de conditionnement qui avait permis d'entreposer les boues de traitement des effluents radioactifs et les cendres d'incinération sur le site de Mol, près de la surface, tenant compte du très faible pouvoir de rétention du sol dans la région de Mol ainsi que de l'existence à Mol d'une nappe aquifère importante qui sert de source d'alimentation en eau potable. En effet, le bétonnage des déchets, tel qu'il avait été appliqué dans les premières années de fonctionnement du C.E.N./S.C.K. nous semblait insuffisant comme conditionnement préalable au stockage, principalement à cause de la vitesse d'élution de la matière, l'hétérogénéité de la matière conditionnée et sa sensibilité aux conditions atmosphériques.

Nous étions donc à la recherche d'une technique permettant d'insolubiliser la matière sous la forme d'une masse homogène.

.../...

Au départ nous avons envisagé de nous servir de matières plastiques, telles que le polyéthylène, qui contiennent couramment des "fillers" minéraux, tels que le graphite ou la craie. Toutefois, à cause du coût des matières plastiques, nous avons finalement opté en faveur du bitume. Une autre possibilité qui s'offrait à ce moment était la vitrification des matières faiblement actives; cette dernière approche fut cependant vite abandonnée par crainte de la complexité de cette technologie; il faut dire que nous sommes revenus à cette idée depuis quelque temps.

Ce souci d'améliorer le conditionnement par le bitumage dans une perspective de stockage des déchets conditionnés près de la surface apparaît d'ailleurs clairement dans une communication faite en 1959 à la Conférence "Ground Disposal of Radioactive Waste" tenue à Berkeley, California.

Je lis :
"The long term storage of chemical sludges resulting from the various
"treatment processes raises the question of fixation so as to avoid
"ground water contamination. One method for insolubilizing sludges
"under investigation is mixing with hot paraffin or tar".

et plus loin :
"By encapsulating radioactive sludges in insoluble petroleum by-pro-
"ducts more economical and safer storage should be possible and long
"term storage at the Mol site may become an acceptable means of solids
"handling".

C'était en 1959.

PREMIERS PAS

Le système du bitumage a été rapidement essayé à l'aide d'un équipement rudimentaire.
Ces essais préliminaires (Fig. 1 à 5) ont cependant permis de sélectionner une approche technologique, notamment le mélange rapide, et de prouver l'intérêt du principe sur base des propriétés de la matière

.../...

obtenue :

- vitesse d'élution d'environ 10^{-6} g.cm^{-2}.j.
- possibilité d'inclure de l'ordre de 50 % des boues sèches.
- parfaite homogénéité de la matière et densité 1.2.

De ce fait il s'est avéré que la qualité du conditionnement recherchée en vue d'un stockage en surface améliorait en même temps les perspectives de rejet en mer. C'est ce qui s'est produit.
Le système développé à Mol consiste donc en l'injection en continu de boues préséchées - 50 % d'humidité - dans du bitume chaud. Un petit "pilote" a été construit et a fonctionné de 1960 à 1964 (Fig. 6, 7, 8). En 1964 l'unité "Mummie" fut mise en marche (Fig. 9) suivant un schéma légèrement amélioré. Cette installation a traité tous les concentrats et toutes les boues produites au C.E.N./S.C.K. depuis 1964 et cela sans incidents et sans problèmes notables.
M. Van de Voorde nous en dira davantage.

SYSTEMES PLUS AVANCES

D'autres systèmes furent développés en plusieurs endroits. Ils sont parfois plus sophistiqués et furent conçus en vue d'applications en des conditions plus difficiles.
Au début des années '60, deux systèmes furent essayés en France; l'un s'appelait LUWA et comprenait comme pièce centrale un évaporateur à couche mince dans lequel la matière desséchée est enrobée de bitume; l'autre est basé sur la presse d'extrusion (Fig. 10, 11). Dans le dernier système les boues humides sont d'abord mélangées à l'émulsion de bitume avec addition d'un produit tensioactif. La déshydratation se fait en grande partie à basse température, par simple séparation des phases; le système fonctionne en continu.
D'autres approches, plus ou moins différentes, furent étudiées dans plusieurs pays d'Europe et aux Etats-Unis. Les Fig. 12 et 13 montrent le système étudié à Oak Ridge. Par ailleurs, lors d'une récente visite en U.R.S.S. nous avons vu une exploitation en parallèle des systèmes à injection directe et la presse d'extrusion.

.../...

La plus grande différence des deux systèmes réside dans le fait que, en cas de traitement d'effluents liquides, la presse d'extrusion permet d'éliminer l'eau et les sels contenus tandis que, en cas d'injection directe, les sels sont absorbés par le bitume. Dans le dernier cas cet inconvénient peut cependant être évité par un prétraitement approprié et filtration des boues.

APPLICATIONS SPECIALES

En parallèle avec les nouveaux développements technologiques le principe du bitumage a été appliqué dans des conditions plus difficiles, ce qui a d'ailleurs permis d'en apprécier les limites d'application. Ces dernières furent résumées par Rodier en 1966 lors d'un congrès à Richland, Washington :

- la température : le bitume devenant mou à partir de 60 ou 70° C, il ne convient pas pour le conditionnement de déchets autochauffants.
- le rayonnement : le bitume étant dégradé à partir d'une dose de 10^9 à 10^{10} rad, la sensibilité au rayonnement constitue une autre raison de réserver l'application du bitume aux déchets autres que les H.L.W.

Alors que le bitumage a été mis au point initialement pour les déchets faiblement actifs (∼ mci/l), il existe déjà une longue expérience dans l'enrobage des déchets moyennement actifs; toutefois il est déconseillé de faire des mélanges déchets/bitume dont l'activité spécifique serait nettement supérieure à 1 Ci/l.

- la teneur en sels solubles du mélange déchets/bitume qui en augmente la vitesse de lixiviation.
- la teneur en nitrates : la présence de nitrates pose certains problèmes de sécurité aussi bien durant l'opération d'enrobage que durant le stockage et cela pour des raisons de danger d'incendie. Ce problème, dont on discute encore aujourd'hui, était reconnu au départ.

  Pour les solutions contenant des quantités importantes de nitrates il y a lieu de conseiller, soit de choisir une autre technique d'enrobage, soit de séparer les nitrates p.ex. par une réaction de précipitation suivie de filtration.

.../...

CONCLUSION

Comme tout système de conditionnement de déchets radioactifs, le bitumage présente des limites d'application. Ces limites sont connues et portent principalement sur l'activité spécifique, la teneur en sels solubles qui augmente le taux de lixiviation et la teneur en nitrates qui, à des concentrations élevées, pourrait créer un certain risque d'incendie.

Néanmoins le système a fait ses preuves et a rendu de grands services. Il permet notamment d'insolubiliser d'un façon efficace un grand nombre de types de déchets radioactifs et, cela avec réduction importante du volume final de la matière conditionnée.

Pour conclure, je voudrais rappeler les conclusions d'une étude de Burns et Clare, publiées en 1968, résumant comme suit les perspectives d'application du bitumage :

- les boues chimiques, virtuellement sans restriction;
- les échangeurs d'ions en faibles quantités;
- les solutions de régénération moyennant des restrictions quant à la teneur en nitrates et sels solubles;
- les cendres d'incinération;
- les matières plastiques.

BIBLIOGRAPHIE

- Ground Disposal of Radioactive Waste - Conference Proceedings Berkeley, California, August 25-27, 1959, p. 37 - 39.
- Research Program on the Treatment and Storage of Radioactive Wastes by L. Baetslé, P. Dejonghe, R. Lopes Cardozo, W. Maes, E.S. Simpson and N. Van de Voorde - 1962; Part II p. 29 - 35.
- Insolubilisation de Concentrats Radioactifs par Enrobage dans de l'Asphalte par P. Dejonghe, N. Van de Voorde, J. Pyck et A. Stijnen. 1964

.../...

- Perspectives d'Emploi du Bitume pour l'Enrobage des Produits Radioactifs par J. Rodier, G. Lefillatre et R. Estournel - Symposium on the Solidification and Long-term Storage of highly Radioactive Wastes - Richland, Washington; February 14-18, 1966.
- Types of Waste suitable for Incorporation into Bitumen by R.H. Burns - G.W. Clare; AERE-M 2144 - 1968.
- Atomic Energy Review - Vol. IX, NO 3 (reprint) 1971 - p. 561 etc.

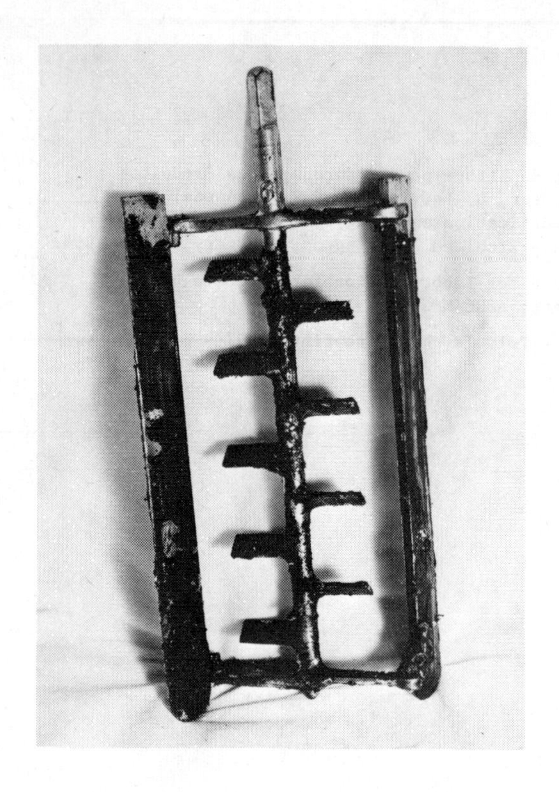

Figure 1

PREMIERS ESSAIS DE MELANGE : MALAXEUR

Figure 2

PREMIERS ESSAIS DE MELANGE : MONTAGE INITIAL

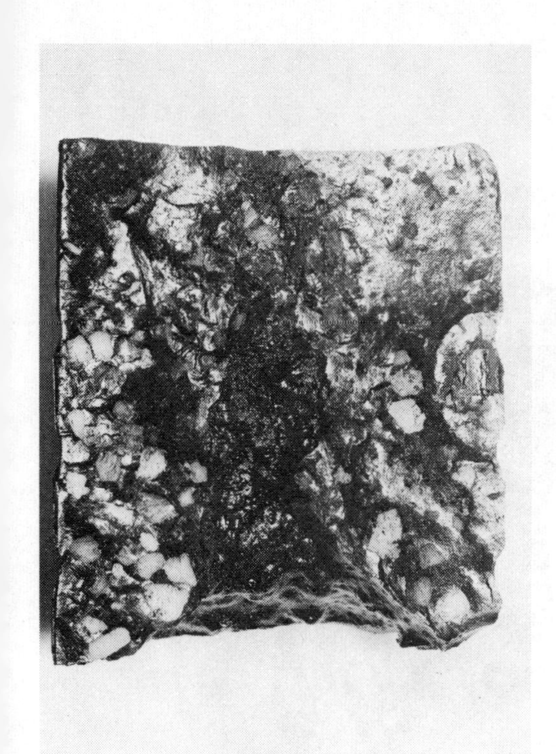

Figure 3

PREMIERS ESSAIS DE MELANGE :
MELANGE DE QUALITE NON ACCEPTABLE

Figure 4

PREMIERS ESSAIS DE MELANGE :
MELANGEUR

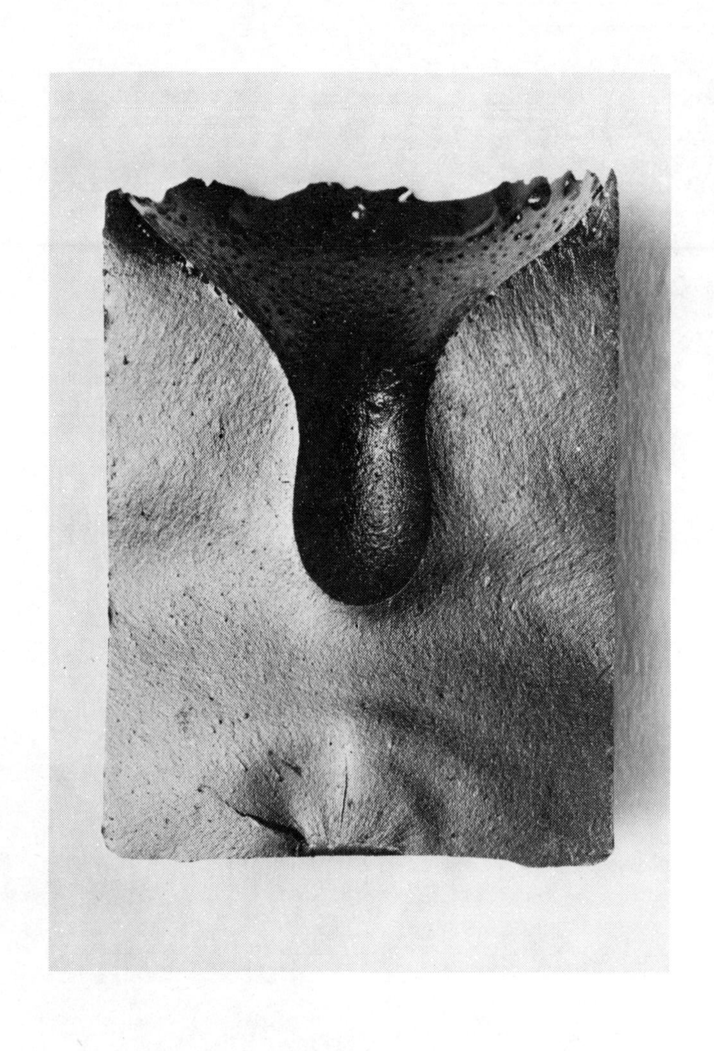

Figure 5

PREMIERS ESSAIS DE MELANGE :
MELANGE DE QUALITE ACCEPTABLE

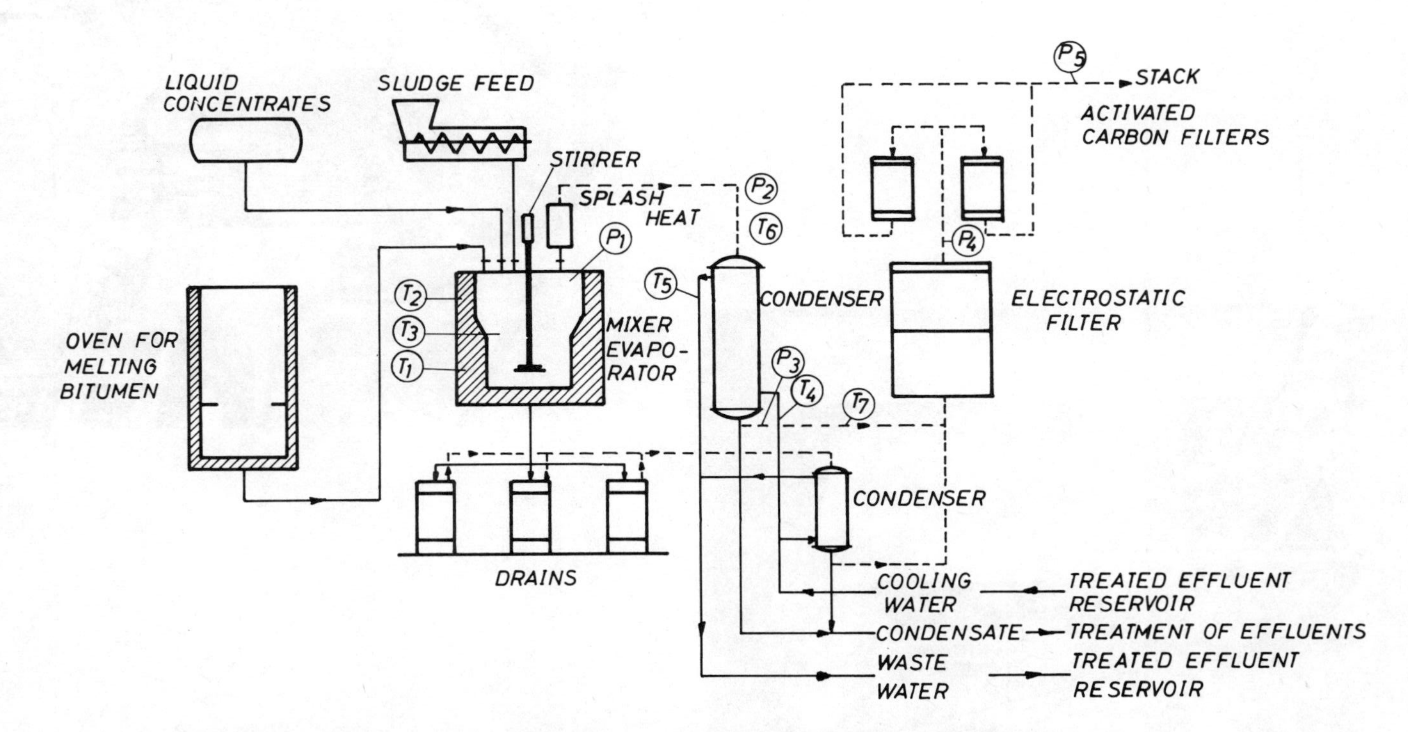

PROCESS FOR FIXING RADIOACTIVE WASTES IN BITUMEN AT MOL

FIG. 6

Figure 7

INSTALLATION PILOTE D'ENROBAGE AU BITUME : VUE D'ENSEMBLE

Figure 8

INSTALLATION PILOTE D'ENROBAGE AU BITUME : SYSTEME D'INJECTION ET DE MELANGE

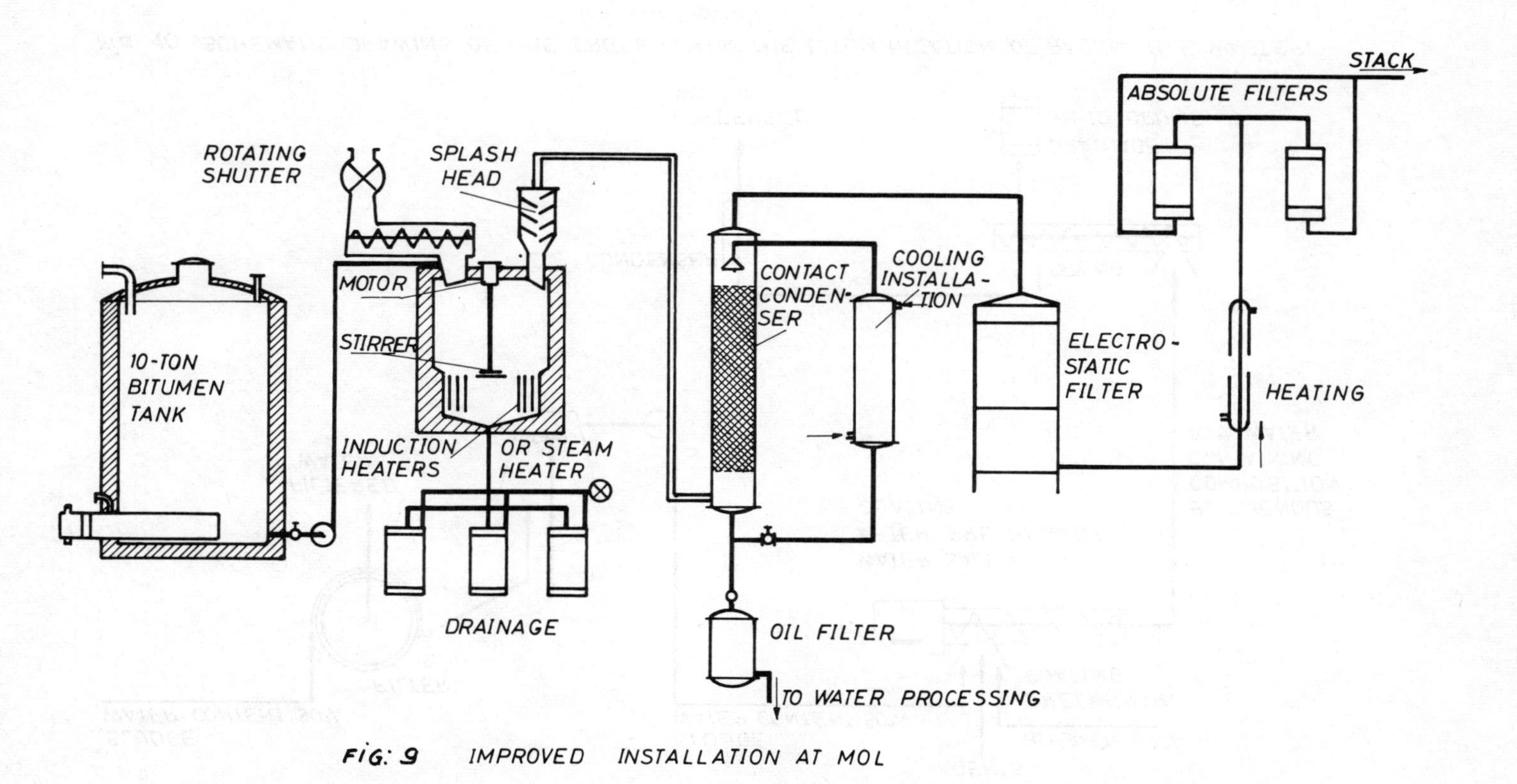

FIG. 9 IMPROVED INSTALLATION AT MOL

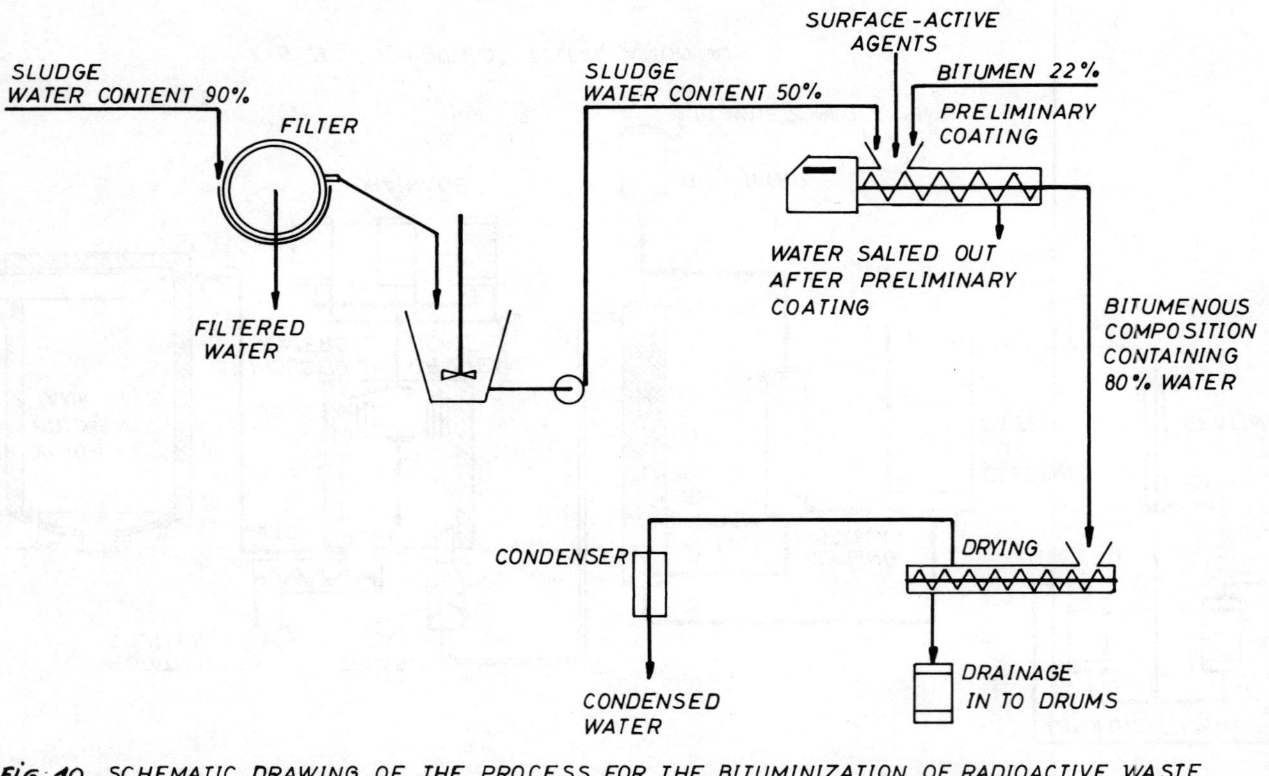

FIG. 10 SCHEMATIC DRAWING OF THE PROCESS FOR THE BITUMINIZATION OF RADIOACTIVE WASTE IN FRANCE

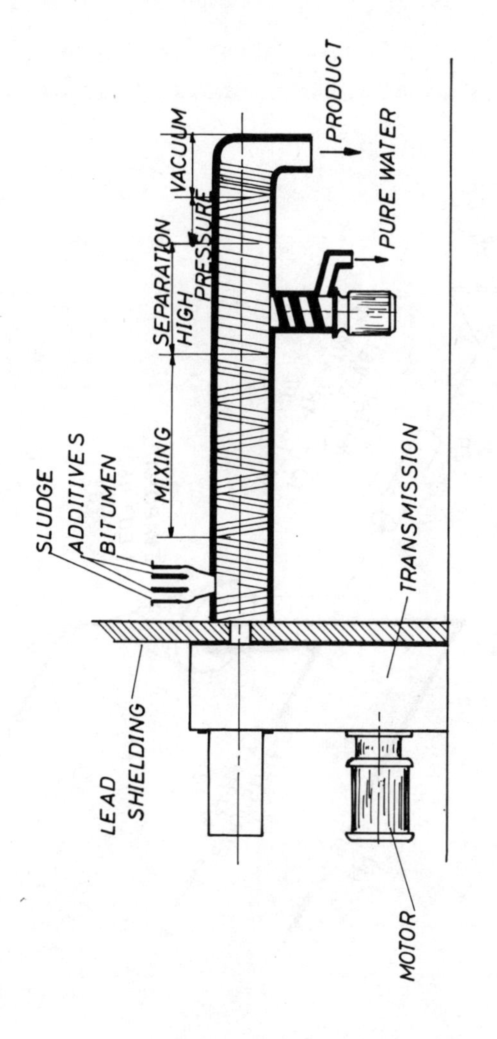

DIAGRAM OF THE COATING MACHINE AT MARCOULE

FIG. 11

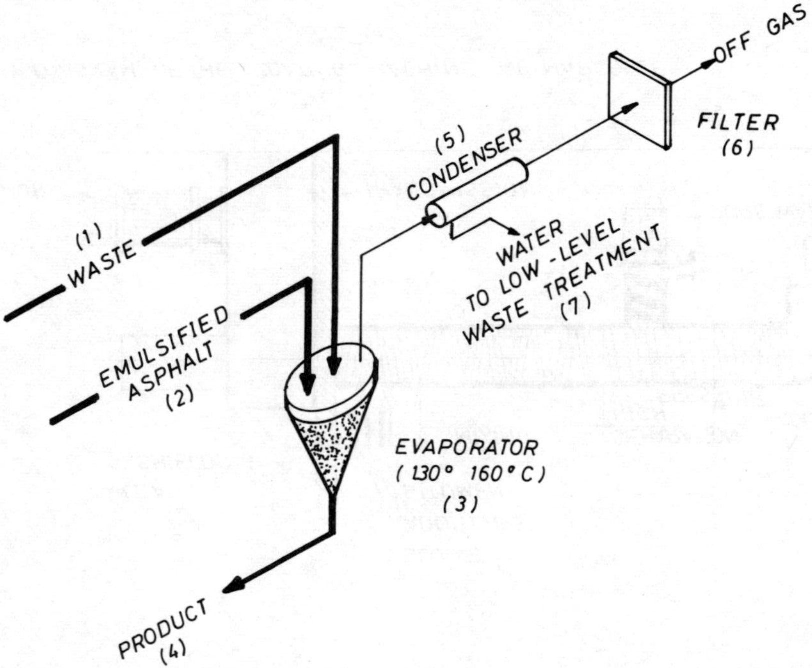

(1) WASTE
EMULSIFIED ASPHALT (2)
EVAPORATOR (130° 160° C) (3)
PRODUCT (4)
(5) CONDENSER
WATER TO LOW-LEVEL WASTE TREATMENT (7)
FILTER (6)
OFF GAS

FIG.12 SCHEMATIC FLOWSHEET FOR FIXATION OF MEDIUM-LEVEL WASTE IN BITUMEN

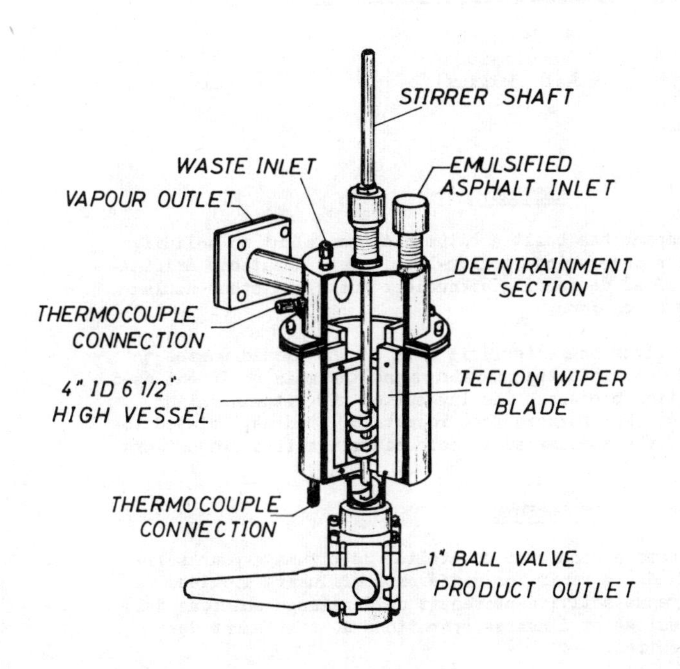

DIAMETRIC DRAWING OF STIRRED EVAPORATOR USED IN LABORATORY STUDIES OF INCORPORATING WASTE IN EMULSIFIED BITUMEN

FIG. 13

THE BITUMINIZATION OF RADIOACTIVE WASTE SOLUTIONS AT EUROCHEMIC

H. Eschrich
Eurochemic
Mol (Belgium)

ABSTRACT

The Eurochemic Company has built a bituminization plant to solidify in the coming years the various medium-level waste solutions originating from the chemical decladding of nuclear fuels and other nuclear fuel reprocessing operations.

The present paper gives characteristic data on the liquid wastes to be solidified and describes the procedures and the main equipment used in the bituminization process. The lay-out of the bituminization plant is presented. Furthermore some results of studies, related to the realization of the medium-level waste bituminization project, are reported.

RESUME

La Société Eurochemic a construit un atelier de bitumage pour solidifier dans les années à venir les différents effluents liquides radioactifs de moyenne activité provenant du dégainage chimique des combustibles nucléaires et d'autres opérations de traitement des combustibles nucléaires.

Ce rapport donne les caractéristiques des déchets liquides qui doivent être solidifiés et décrit le mode opératoire et l'équipement principal employé pour l'opération de bitumage. La description de l'atelier de bitumage est présentée. En outre, quelques résultats d'études relatives à la réalisation du projet de bitumage des déchets radioactifs de moyenne activité, sont rapportés.

1. HISTORICAL AND TECHNICAL INTRODUCTION

The European Company for the Chemical Processing of Irradiated Fuels (EUROCHEMIC) is a joint undertaking of the OECD Nuclear Energy Agency (NEA). EUROCHEMIC was legally constituted as an international shareholding company in July 1959.

Member countries are : Austria, Belgium, Denmark, France, Federal Republic of Germany, Italy,(The Netherlands, until 1975), Norway, Portugal, Spain, Sweden, Switzerland, and Turkey.

The objectives of the Eurochemic Company have been :

- to carry out research or industrial activities connected with the chemical processing of irradiated fuels,
- to meet reprocessing demands of member countries, and
- to train specialists in the field of nuclear fuel reprocessing.

For these purposes a reprocessing plant and a research laboratory were built and operated.

The construction of the plant and the auxiliary buildings started in July 1960 and has been completed in 1964.

The research laboratory - consisting of a cold wing, hot laboratories, hot cells and an engineering hall - could be availed already at the end of 1963.

Inactive tests on the plant equipment were made in 1964 and active tests during the first half of 1966. The active operation of the plant started in July 1966.

In 1971 the Board of Directors of the Company decided to terminate the fuel reprocessing activities at the end of 1974 and to commence the conditioning of the wastes generated and stored on-site.

The Eurochemic installations allowed the successful processing of irradiated fuels from research, material testing, and power reactors of widely different characteristics as far as enrichment, dimensions, burn-up, core and cladding material are concerned.

The most outstanding features of the applied chemical process schemes have been :

- the chemical decladding of the fuel elements,
- a two-cycle Purex process for enrichments of up to 5 % in U-235 using 30 % tributyl phosphate (TBP), and
- a three-cycle TBP-process for initial enrichments of up to 93 % in U-235 using 5 % TBP in the first two cycles and 30 % TBP in the third cycle.

Considering the present situation of the Eurochemic Company and in view of a safer storage, the Belgian Authorities have requested to convert all high-level waste (HLW) and medium-level waste (MLW) solutions - originating from the past reprocessing and the present decontamination activities - into solids within a reasonable period of time.

During the approximately eight years of active operation, Eurochemic has processed 181 tons of low-enriched and 30.6 tons of high-enriched uranium fuels and has produced - besides various kinds of solid wastes and low-level liquid wastes - about 870 m^3 high-level and about 2000 m^3 medium-level aqueous waste solutions.

Up till now the entire low-level liquid wastes and most of the solid wastes are sent to the neighbouring CEN - BelgoNucléaire waste treatment installations for further conditioning whilst the high- and medium-level waste effluents have been stored in tanks on the Eurochemic site.

At the present time the active part of the plant equipment is thoroughly decontaminated which will most probably increase the volume of the solutions belonging to the medium-level waste category (0.1 - 10000 Ci/m^3) by some hundred cubic meters.

When Eurochemic decided to apply chemical decladding procedures it was fully aware of the consequences with respect to the management of the resultant waste solutions. Therefore research and development work on the solidification of the various decladding solutions started already in 1960. In most of the R and D work that followed the possibility of treating the high-level fission product solutions from the processing of MTR-fuels together with the decladding solutions has been considered.

Careful evaluation of existing solidification processes, own development work and tests on industrial scale and furthermore the favourable results obtained by the bituminization techniques in several countries, notably in Belgium, France and the Federal Republic of Germany, led Eurochemic to the decision to solidify its MLW solutions by incorporation into bitumen.

The present paper gives :

- relevant information on the waste solutions to be solidified,
- a description of the chemical processes and the operations involved in the bituminization process,
- a description of the lay-out and the equipment of the Eurobitum plant, and
- information on some R and D studies related to the realization of the MLW-bituminization project at Eurochemic.

2. THE BITUMINIZATION OF MEDIUM-LEVEL WASTE SOLUTIONS AT EUROCHEMIC

2.1 Origin, Composition and Amounts

At Eurochemic the aqueous MLW solutions originate from three basically different sources :

- from the chemical decladding of fuel elements,
- from the decontamination of active plant equipment, and
- from the concentration of a certain type (hot waste) of a low-level waste solution, the socalled hot waste concentrate (HWC).

The decladding waste comprises four types according to the procedure used to dissolve the different cladding material of the fuel :

- Zirflex waste from Zircaloy-clad fuel,
- Sulfex waste from stainless steel-clad fuel,
- aluminium waste from aluminium-clad fuel, and
- magnesium waste from magnesium alloy-clad fuel.

The special decontamination solutions consist of alkaline and acidic solutions containing mainly oxidants and complexing agents.

The HWC is produced from that fraction of low-level waste which has a higher specific activity than 3×10^{-2} Ci/m^3. This fraction, - which can also include large volumes of simple decontamination solutions (e.g. HNO_3 and NaOH), - is concentrated in an evaporator until the salt (mainly sodium nitrate) concentration approaches the saturation limit. No decladding waste originates from the processing of highly-enriched fuel elements (MTR-fuels) as the canning and core materials are simultaneously dissolved.

The radionuclides present in the decladding effluents come from the neutron activation of the cladding material constituents and from losses of the irradiated fuel cores (fission products and actinides); these losses of core material to the decladding wastes have been found to be less than 1 % for all dissolution procedures applied.

Some principal data on the composition and amounts of Eurochemic's MLW solutions are given in Table I.
The production of decladding solutions is terminated also in the case reprocessing activities would recommence as then - as already foreseen for many years - a chop and leach head-end would be used. The final volume of decontamination wastes is difficult to predict; until May 1976 about 100 m^3 were produced and mixed with either the aluminium waste or the hot waste concentrate.

TABLE I - DATA ON EUROCHEMIC'S MLW SOLUTIONS

(JDW = Jacket Decladding Waste)
(SS = Stainless steel)

WASTE SOL.	CONCENTRATION OF MAIN COMPONENTS	SPEC. ACTIVITY (May 1976) Ci/m^3	VOLUME (May 1976) m^3
Al-JDW	2 M $NaAlO_2$, 2 M NaOH	< 400	150
SS-JDW	0.8 M SS-SO_4 , 2 M H_2SO_4	1000	135
Mg-JDW SS-JDW	0.58 M SS-SO_4 , 0.21 M Mg SO_4 0.1 M Mo , 1.8 M H_2SO_4	800	260
Zr-JDW	0.4 M Zr , 2.7 M F^- 1.1 M NH_4 , 0.1 M NO_3^-	800	400
HWC	5 M $NaNO_3$, 2 M HNO_3 (NH_4^+ , Mn, organic complexing agents present)	700	1160

The MLW volumes produced per ton of fuel processed are given in Table II.

The maximum specific activity of the MLW solutions after several years storage was found in the Sulfex waste and amounts to about 1 Ci/l.

The MLW solutions are stored in two buildings, (Bldg. 21 and 24) of which one houses six 260 m^3 tanks and the other four 500 m^3 tanks.

The general implantation of the LLW and MLW storage and treatment buildings is shown in Fig. 1.

TABLE II - SPECIFIC MLW PRODUCTION AT EUROCHEMIC
(m^3 waste/ton U fuel)

WASTE SOL.	Al-JDW	SS-JDW	Mg-JDW	Zr-JDW	HWC
VOLUME	2.1	4.5	2.6	5.4	5

As far as acceptable, different MLW solutions were mixed to make maximum use of the installed storage capacity. Thus, Sulfex solution was mixed with the magnesium decladding solution (a weak sulfuric acid solution of the magnesium cladding), and decontamination solutions were appropriately mixed with the aluminium decladding solution and hot waste concentrates. Arbitrary mixing of MLW solutions would lead to the formation of precipitates and/or corrosive solutions.

2.2 The Bituminization Process

The Eurochemic MLW bituminization process consists of two essential steps :

- the batchwise chemical pretreatment operations, and
- the continuous incorporation of the resulting slurries into bitumen using a screw-extruder-evaporator.

The purpose of the chemical pretreatment is :

- to insolubilize to a high degree the radionuclides contained in the MLW effluents by precipitation and coprecipitation (mainly hydroxides, barium and calcium sulfate, calcium phosphate, calcium fluoride, nickel ferrocyanide),
- to minimize the corrosiveness of mixed waste solutions towards equipment materials, and
- to eliminate ammonium (which otherwise would represent a potential risk during the bituminization operation) by boiling the alkaline slurry.

The waste slurries obtained by the chemical pretreatments consist of about 40 wt.-% salts and 60 wt.-% water.

These slurries are mixed together with molten bitumen in the extruder where the water is evaporated and the remaining solids are homogeneously distributed within the bitumen matrix at temperatures below 200 °C.

The hot bitumen-waste product, leaving the extruder, is cast into 220 l "chromized steel" drums; it is then allowed to solidify and to cool down before it is transferred to concrete bunkers for an interim storage on the Eurochemic site ("Eurostorage" facility).

The bitumen-waste product in one drum amounts to about 180 l (or about 245 kg) and consists of 45 wt.-% solids, 55 wt.-% bitumen, and less than 0.5 wt.-% water; the specific activity will be maximum 1 Ci/l (or about 180 Ci/drum resulting in a dose rate of about 220 Roentgen/h at the drum surface).

The MLW solidification process will be realized in the Eurobitum-installation described hereafter. An example of a chemical pretreatment flowsheet and a slurry incorporation flowsheet are given in Fig. 2 and Fig. 3 respectively.

The MLW solidification capacity is determined by the evaporation rate of the extruder-evaporator and the operation time.

Assuming a three-shift schedule and an evaporation rate of 140 kg water per hour the yearly throughput will be 650 m^3 MLW corresponding to the production of 3.600 bitumen-waste drums per year.

2.3 Description of the Engineering Flowsheet and the Main Equipment

In Fig. 4 a simplified functional flowsheet is given. It shows the chemical pretreatment, the slurry bituminization, and the off-gas treatment system.

2.3.1 Waste transfer system

The liquid waste solutions are transferred by steam jet from building 24 to the diverter 1 in building 26 via two pipes in a shielded aerial transfer-line.

A third pipe is foreseen in this transfer-line for returning effluents from drain tank 6 to building 24.

Diverter 1 (material : D W N 4505) has five outlets and two spare ones for later extension.

2.3.2 Waste buffer tanks

Four air-sparged buffer tanks of 9.2 m^3 each receive the waste solutions from the diverter. A level-order-high stops the liquid transfers from building 24.

From the buffer tanks 2, 3, 4, and 5 the measured amounts of waste solutions are transferred by jets to the reaction vessels 7 or 8.

The buffer tanks are equipped with an overflow to the drain and rework tank 6.

2.3.3 Drain and rework tank

Vessel 6 acts as a rework vessel; it has a useful volume of 1.7 m^3 and can receive solutions as well as discharge them to the vessels indicated in Fig. 4.

2.3.4 Reaction vessels for chemical pretreatments

The two reaction vessels 7 and 8 are identical; each has a volume of 3 m^3 and is equipped with three jackets for heating or cooling purposes. The vessels are furthermore equipped with :

- propeller-type stirrers,
- a sampling system,
- temperature, density and level measuring systems,
- connections to the reagent make-up vessels 14, 15, 16 and 17, to receive liquid chemicals,
- connections to two weighing bins, to receive chemicals in powder form from silos 11 and 12,

- a steam-heated off-gas line connecting the vessels 7 and 8 either to the middle of the ammonia-elimination column 20 or to the general vessel ventilation scrubber 22; a droplet separator 18 is installed in the ammonia ventilation line,

- lines to an empty cell, which can, in case of necessity later on, receive a liquid-solid separation system 9.

The slurries formed in the vessels 7 and 8 are routed to the slurry feed-tank 10 via self-proming pumps.

2.3.5 Liquid reagents make-up unit

The cold make-up system for delivery of liquid chemicals consists of four 1 m^3 cylindrical stirred tanks 14, 15, 16 and 17 made of stainless steel.

2.3.6 Solid reagent silos and metering unit

Two silos 11 and 12, receive respectively calcium hydroxide and barium hydroxide in powder-form. Calcium hydroxide is delivered to the site by truck from which it is transferred pneumatically to its silo.

A hopper system at the silos assures continuous feeding of the weighing bin 13 from where the powdery reagents are transferred by gravity to the reaction vessels 7 and 8.

2.3.7 Ammonium elimination system

When ammonia is boiled off in the tanks 7 or 8 the steam-heated off-gas line of these tanks is connected to the middle of the steam-heated column 20. Most of the water vapours are condensed while the gaseous ammonia is routed, after sufficient dilution with hot air (heater 24), through filter 93 to the stack.

The course of the ammonia removal is followed by means of conductivity monitoring of the condensate recycled to the column 20.

The condensate of column 20 is collected in vessel 21 from where it can be sent to the hot waste tank 31 or to a warm waste tank in the low-level waste treatment building 8 (see also Fig. 1).

When no ammonia must be eliminated from vessels 7 or 8 they are no longer connected to the ammonia separation column 20 but to the general vessel ventilation line.

2.3.8 Vessel ventilation system

The gases coming from the vessels pass successively through a caustic scrubber 22 which can be heated, a heater, and a filter unit 25 consisting of a prefilter and an absolute filter.

A diagramme of the vessel ventilation and the off-gas (ammonium elimination) system is shown in Fig. 5.

2.3.9 Slurry feeding system

The slurry feed tank 10 has a volume of 7 m^3; this capacity corresponds to 1.5 day operation.

The tank is made of AISI 304 L and is equipped with :

- slurry inlet connections from vessels 7 and 8,

- a stirrer to keep the slurry homogeneous,

- a double jacket, for heating or cooling,
- maximum and minimum level alarms,
- measurement systems for temperature and density,
- a sampling system,
- connections to the make-up vessels for special after-treatments,
- connections to the extruder via self-priming metering pumps.

2.3.10 Bituminization unit

The slurry is pumped from tank 10 to extruder 27 where it is mixed with hot bitumen.

2.3.10.1 The extruder

A VDS-V 83 extruder (manufactured by Werner u. Pfleiderer, Stuttgart) has been installed for the incorporation of the slurry in molten bitumen. The extruder 27 evaporates the water of the slurry at a rate of about 140 l/h and incorporates homogeneously the remaining solids in the bitumen.

The extruder is jacket-heated by saturated steam of 20 kg/cm^2 produced in unit 35. Three separate heating zones are foreseen along the length of the extruder (see Fig. 3).

The inlet part is connected to the slurry and molten bitumen feed lines and two decontamination lines, one for an organic solvent (e.g. tetrachloroethylene), the other for water.

A steam-heated dome (chimney) with inspection window, surmounts each of the three heating sections of the extruder. Along these domes the water vapours are routed to separate condensers.

2.3.10.2 Distillate routing and oil removal

The aqueous condensate, collected in tank 28 at the outlet of the condensers of the extruder, contains up to 0.05 % bituminous oil which is separated from the water on the filter 29. The spent filters loaded with the oil are dumped into bitumen end-product drums and coated with bitumen.

The aqueous phase is either routed to the hot waste vessel 31 or to the warm waste drain tank in Bldg. 8, depending on its activity which is measured by an inline gamma monitor.

2.3.10.3 Filling station and drum handling

The installation for filling and handling the final waste product drums comprises :

- a turntable 37 bearing 6 removable drip-trays, each tray can take up a drum of 220 l,
- a steam-heated funnel provided with a valve which collects the fluid product during drum changing,
- a powerful ventilation to remove rapidly all vapours from the product drum beneath the extruder outlet,
- conveyors for empty and filled drums,

- hydraulic grab tools to put empty drums on the turntable and filled ones from the turntable to a conveyor and from there to a waggon which transports each time twelve drums to the storage bunkers,

- product level measuring devices,

- shielding doors separating the different cells,

- a drum closing device,

- a small waggon for removing drums from the filling station, and

- a telemanipulator which serves for interventions in the drum handling cells.

Decontamination solutions are routed to the evaporator 46 to recover the organic solvent in tank 33 while the bitumen residue is poured from the evaporator into a waste container.

2.3.10.4 Bitumen tank and feeding system

The molten bitumen arrives in a tank-truck on-site from where it is transferred to the mild steel tank 34 of 30 m^3 volume. The steam-heated tank is located outside the building at its East side.
From the tank the molten bitumen is fed to the extruder by means of a heated dosating pump and through steam heated pipes, assuring a transfer temperature of about 165 °C.

2.3.11 Control room

This room has four lead-glass windows to allow observation of all operations.

It contains :

- access to the driving mechanism of the turntable,

- interlocks and controls for the various operations required for the handling of empty and full drums,

- the control panel for the process.

2.3.12 Analytical process and product control

The waste solutions in the buffer tanks 2, 3, 4, 5 and the slurries from the chemical pretreatments in tanks 7, 8, and 10 can be sampled by a Thorex-type system in a shielded blister; the tanks 21, 23, 30, and 31 containing low active solutions can be sampled in an unshielded blister. Some of the most essential analyses of the slurries are carried out directly in the sampling blister whilst all other analyses on solutions are performed in an analytical box which receives the samples by a pneumatic post.

The bitumen waste product can be sampled in the filling station and subsequently be analyzed using a sampling device and analytical methods developed by Eurochemic.

2.4 Bituminization Building and Lay-Out

The main equipment described in the previous chapters and nearly all the auxiliary equipment are housed in the bituminization building (Bldg. 26, see Fig. 1) annexed to the effluent treatment building 8.

The bituminization building is approximately 38 m long, 25 m wide and 10 m high and has three floors and about 50 rooms or cells. It is constructed in reinforced concrete.

The ground floor (Fig. 6) houses the active main process equipment, such as the MLW waste buffer vessels, reaction vessels, slurry feed vessel, extruder, filling station and the drum handling devices.

The control room, steam generator, solvent evaporator, drum closing device and the electrical sub-station are also situated on the ground floor.

The first floor (Fig. 7) houses the auxiliary equipment, such as the sampling blisters, reagent make-up vessels, valve- and transmitter-galleries, vessel ventilation and cell and room ventilation fans and filters.

The second floor houses the cooling water facilities and the storage of solid reagents and their transfer equipment.
(A large storage silo for calcium hydroxide is placed outside the building from where the powdery reagent is pneumatically conveyed into the solid reagents room).

Construction of the bituminization plant has started at the end of 1972.

The commissioning tests commenced in October 1974 and the active start-up is scheduled for the second half of 1976.

3. DEVELOPMENT WORK RELATED TO THE MLW BITUMINIZATION AT EUROCHEMIC

The past development work on MLW, which formed the basis for Eurochemic's bituminization plant project, has been described in a previous paper [1]. In the present chapter a brief description of some recent studies related to the realization of the bituminization plant project is given.

3.1 Chemical Pretreatments

Eurochemic has developed suitable chemical pretreatment flowsheets for the bituminization of the combined MLW-HEWC (High-Enriched Waste Concentrate from MTR-fuel reprocessing) solutions resulting in a bitumen waste product of about 4 Ci/l during the incorporation period [1].

A drawback of the developed pretreatment processes appeared to be the use of large amounts of expensive barium hydroxide powder. About 50 % of the costs of added chemicals would have been due to this reagent.
It has recently been requested that the specific activity of the final bitumen waste products should not exceed 1 Ci/l, thus, the addition of larger amounts of HEWC (about 20 Ci/l) is no longer possible.
Consequently the established and tested pretreatment flowsheets - which involved a mixing of all waste types - are not applicable anymore.

Extensive studies were carried out aiming at an optimization of the chemical, physical and economical conditions of MLW pretreatment processes guaranteeing a safe incorporation of the resulting slurry and a homogeneous bitumen waste product of low leaching and burning rate.

Besides modifications of the former standard flowsheets new pretreatments have been developed in which the addition of barium hydroxide is partially or completely eliminated. In Fig. 2 an example is given in which it is essential that the phosphate ions are added after the free sulfuric acid of the Mg-SS-JDW has been neutralized.

The new flowsheets have been developed for the simultaneous bituminization of the different MLW solutions.

If the HWC effluent could be treated in another way, a separate treatment of each MLW type would certainly be advantageous. Flowsheets for the treatment of the Zirflex, Sulfex and aluminium decladding solutions have been designed and successfully cold-tested in laboratory scale equipment.

3.2 Treatment of Hot Waste Concentrate (HWC)

The development work aims at a decontamination of the HWC from all alpha-emitting nuclides and long-lived fission products in order to eliminate this type of MLW effluents from its bituminization, to reduce the overall costs of the MLW management and to improve the characteristics of the remaining bitumen waste products (e.g. lower leaching and burning rate). Eurochemic's HWC consists essentially of an almost saturated sodium nitrate solution in nitric acid containing also organic compounds and ammonium nitrate.

It is investigated whether the HWC, as stored, can be decontaminated by simply passing it through chromatographic columns. The column materials studied comprise inorganic ion exchangers, preformed precipitates incorporated or deposited on granular materials and extractant-loaded supports (extraction chromatography).

The results obtained until now are very promising and show that the indicated treatment approach is principally feasible. Thus, passage of two hundred bed volumes HWC-solution through a column composed of titanium phosphate and TBP-coated macroporous polystyrene-divinylbenzene particles resulted in an effluent that was decontaminated in Cs, U and Pu by a factor of $\geq 10^5$. The search for suitable sorbents effectively removing strontium and ruthenium from the HWC is continuing.

The highly-active column materials will be subject to HLW solidification and the column effluent will be treated as LLW.

Moreover, a two-column-system has been developed permitting a quantitative removal of U, Np, Pu, Am and Cm from the HWC after an appropriate pretreatment involving addition of a complexing agent and adjustment of the acidity.

3.3 Hazards caused by Incorporating $NaNO_3$ and $NaNO_2$ into Bitumen

The investigations on this subject and the first results obtained have been reported in a previous paper [1]. In continuation of this work homogeneous bitumen products were prepared containing 40 - 50 wt.-% Mexphalt R 90/40 or Mexphalt R 85/40, 16 - 30 wt.-% $NaNO_3$, 4 - 10 wt.-% $NaNO_2$ and 10 - 40 wt.-% insoluble solids. These compositions cover a wider range than the bitumen products to be expected according to the established Eurochemic flowsheets.

It was found by means of thermogravimetric and differential thermogravimetric analyses that significant exothermic reactions for all investigated compositions occurred exclusively in the temperature range 390 - 430 °C (i.e. beyond the melting points of $NaNO_3$ and $NaNO_2$). Though the temperature, $T_{ex.}$, at which an exothermic reaction (ignition, explosion) occurred is relatively little influenced with respect to the largely different compositions, one can draw the following conclusion : the thermal stability of bitumen/salt mixtures increases if the $NaNO_3$ and/or $NaNO_2$ content is decreased and the content of bitumen and/or insoluble solids is increased.

A great number of experiments were performed to establish safe incorporation conditions.

The evaluation of the differential thermogrammes obtained on differently composed bitumen products (40 - 50 % bitumen + $NaNO_3$ + $NaNO_2$ + insoluble solids) revealed that no exothermic reaction takes place below 295 °C.

An interesting observation has been the decrease of $T_{ex.}$ of 10 - 20 °C upon aging of some bitumen products.

The influence of the particle size of the homogeneously dispersed salts in the bitumen on $T_{ex.}$ was investigated in the range 40 - 200 μ m using samples composed of 48.6 % Mexphalt R 90/40, 23.4 % insoluble solids and 28.0 % $NaNO_3$.

The results led to the conclusion that the smaller the particle size of the dispersed solids the lower the temperature at which a significant exothermic reaction was measured; these temperatures ranged from 400 - 450 °C for the indicated composition.

From the investigations performed one can draw the following main conclusions :

- there exists no risk of a spontaneous exothermic reaction during the incorporation of Eurochemic waste slurries into blown bitumen (Mexphalt R 90/40 or R 85/40) provided that the bitumen/salt mixture contains at least 40 wt.-% bitumen, that the different salt constituents are homogeneously dispersed and that the temperature at any point of the mixture is kept below 280 °C;
- if the homogeneous dispersion of the salt components cannot be assured, a minimum content of 50 wt.-% bitumen is required and the temperature at any point of the mixture should never exceed 230 °C in order to avoid hazardous exothermic reactions.

3.4 Elimination of Ammonium and its Control

Ammonium present in the Zirflex solution and the HWC will be removed as ammonia by boiling the alkaline slurry. It has been proposed to separate the mixture of water vapours and ammonia in a heated packed column which is connected to a reflux-cooler. The degree of the ammonium removal can be controlled in the off-gas line or other parts of the installation.

Laboratory units have been built and tested to evaluate the design of a suitable separation column and the conditions for the most efficient separation of ammonia from water. The most reliable and sufficiently sensitive continuous control of the removal of ammonia was found to be the measurement of the conductivity of the condensate flowing back from the cooler to the separation column.

For the range of 0.01 $\underline{M}$ to 0.88 $\underline{M}$ NH_4OH a relative standard deviation of 0.7 % was found.

The ammonium contents of solutions contained in vessels are analyzed after sampling in an analytical box using the Kjeldahl method for higher concentrations and an ammonium-ion selective electrode for low concentrations.

The completeness of the ammonium removal from the reaction vessels will be controlled by taking a slurry sample and analyzing it - after separation of the precipitates - in the mother liquid by means of the Kjeldahl procedure.

3.5 Density Measurements of Waste Slurries

The applicability of a dip-tube system to the measurement of slurry densities between 1.2 and 1.4 has been tested.

It could be concluded that the dip-tube measuring system as now employed for solutions in the reprocessing plant is equally well-suited for the slurries obtained in the chemical pretreatment of the MLW mixtures.

3.6 pH-Measurements in Radioactive Slurries

A strict control of the pH of the waste solution mixtures during their pretreatment is essential.

A four weeks lasting test with an industrial electrode assembly was carried out. The electrodes were in contact with all solution and slurry compositions occurring during typical chemical pretreatments. The temperature varied between 20 °C and 105 °C when the pH was about 8. The total slurry volume was 0.5 m^3.

After the pretreatment operations the electrode assembly remained for nine months in the slurry at room temperature and the pH was continuously recorded.

The test showed that accurate pH measurements in the expected waste slurries can be performed and that the characteristics of the electrodes do not change noticeably within nine months. It would therefore be possible to perform "in-line" pH-measurements. The influence of radiation on indicator and reference electrodes has been studied (- in cooperation with the Danish Research Establishment at Risö -) applying a total dose of 5×10^6 rads in a period of 90 hours. Apart from the fact that the electrodes became brownish coloured no essential change in their normal properties could be found.

In spite of the positive results obtained it was finally concluded that it is more convenient and economical to control the pH in slurry samples taken from the respective vessels.

3.7 Removal of Bituminous Oil

A great number of organic and inorganic granular materials have been tested for their applicability as filter materials to remove the bituminous oil entrained in the aqueous distillate from the extruder-evaporator. The best material, with respect to flow-resistance and absorption capacity, was found to be an inorganic glassy material of volcanic origin which has been thermally expanded and then chemically treated to render it hydrophobe. This material "EKOPERL K-33" manufactured in FRG showed an average absorption capacity of 200 g bituminous oil per litre filter bed at a linear liquid flow of 0.4 - 18 m/h. The filtrate is clear and completely free from entrained oil droplets.

3.8 Corrosion Tests on the Screw-Extruder Material

Thorough corrosion tests were performed on the screw components made of nitrided mild steel. It was found that under normal operation conditions (pH of slurry about 8-9) no risk of a serious corrosion exists. However, maloperations and special decontamination reagents may damage the extruder material.

A strong corrosion with rapid and complete dislocation of the nitrided surfaces has been observed if the mother liquor of the slurry has a pH ≤ 3.

Corrosion tests over 740 hours with a boiling mixture of bitumen-water-organic solvent (tri- or tetrachlorethylene) showed that the extruder material is attacked, especially in the vapour phase. Depending on the composition of the mixture corrosion rates of 0.03 - 0.05 $g/m^2/h$ were measured.

3.9 Selection of the Container Material for the Bitumen-Waste Products

Originally painted mild steel drums were foreseen as containers for the bitumen waste products. Depending on the storage conditions, these drums corrode more or less fast within some years which may cause contamination of the storage containment and difficulties for the transport from the interim to the final storage place. The corrosion studies performed aimed at an improvement of the corrosion resistance of the steel drums or - if it is shown to be more economical - the selection of a suitable container material more corrosion-resistant than painted mild steel. Thus, the properties of a new type of steel plates recently developed by Cockerill (Belgium), which is said to be corrosion- resistant and to have a surface layer with the properties of a 20 % chromium ferritic stainless steel (the protective coating thickness is 80 - 110 µ with a minimum chromium content of 20 % at the surface and of 13 % at the interface). This chromized iron-plate can be welded by all conventional processes.

Galvanized mild steel has also been taken into consideration for this corrosion study.

The provisional conclusions of the study, which is still continued, converge on the very good behaviour and excellent corrosion stability of the chromized mild steel. The corrosion resistance of this material towards the various corrosive media studied can be compared to that of a 18/8 type stainless steel.

If the price of a standard painted mild steel drum is assumed to be equal to 100 in an arbitrary unit (approximately 400 BF), the respective prices for galvanized steel, chromized steel and stainless steel (of the same shape and dimensions) will be 160, 200 - 225 and more than 900, in the same monetary unit.

It has therefore been considered advantageous to choose the chromized steel material for the drums in which the bitumen-waste products have to be stored.

3.10 Analytical Controls

To guarantee a safe bituminization of the wide variety of waste solutions and the required quality of the solid products, a thorough analytical control of the entire process is necessary. Suitable methods and equipment for the remote sampling and analysis of the medium-level radioactive waste solutions, slurries and bitumen-salt mixtures have been developed and cold-tested.

The analyses to be performed include the determination of the:

- composition of the original waste solutions;

- specific density, salt content (soluble and insoluble), ammonium and nitrate concentrations, radioactivity and pH of the waste slurries;

- quality of the bitumen used;

- pH, activity, ammonium and total salt content of low-level process streams (secondary wastes);

- water content, specific activity, weight fractions of the salts and bitumen in the final solid waste product.

REFERENCES

[1] Hild, W., Detilleux, E., Eschrich, H., Lefillatre, G., and Tits, E. : "Development work on the homogeneous incorporation into bitumen of intermediate-level waste from the Eurochemic reprocessing plant, Management of low- and intermediate-level radioactive wastes", Proc. Symp., Aix-en Provenoe, IAEA, Vienna (1970) - 689.

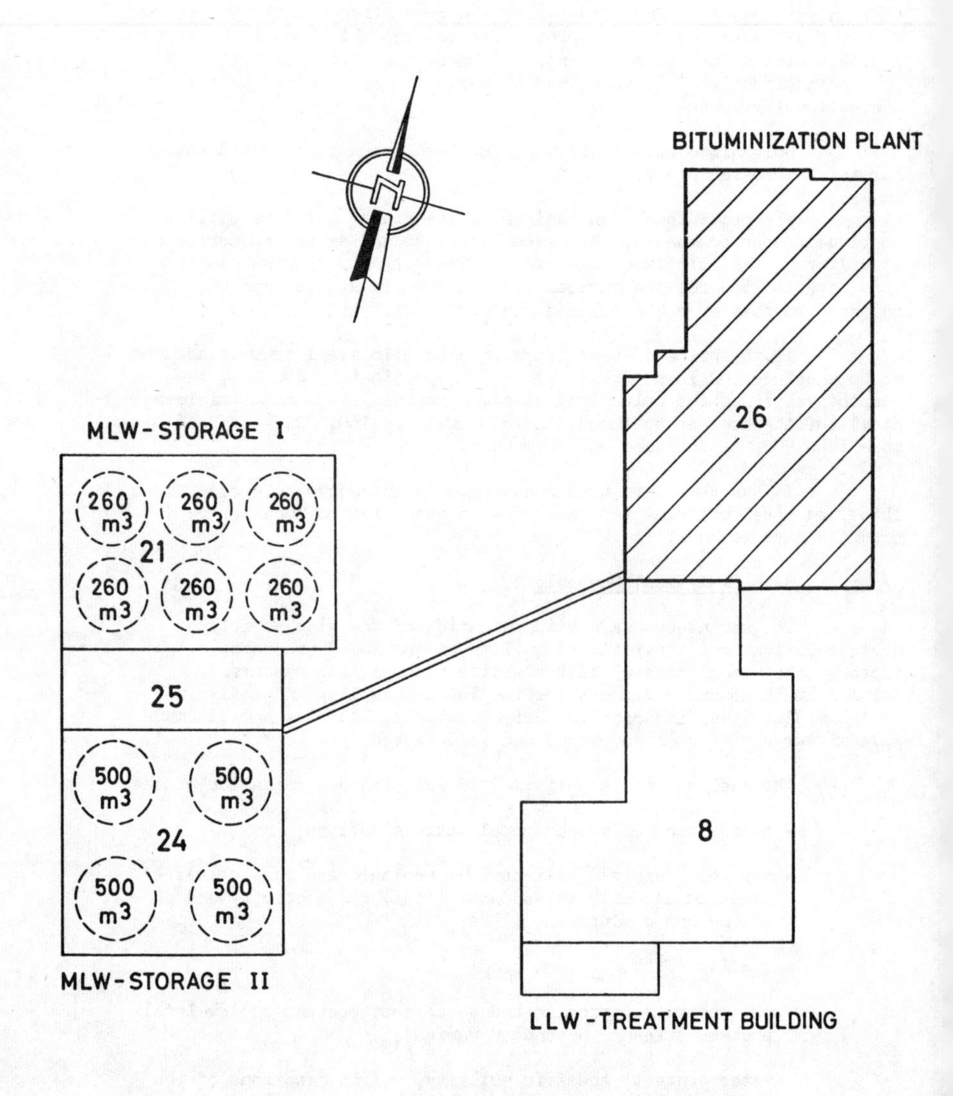

Fig.1 GENERAL IMPLANTATION

Scale ~1: 650

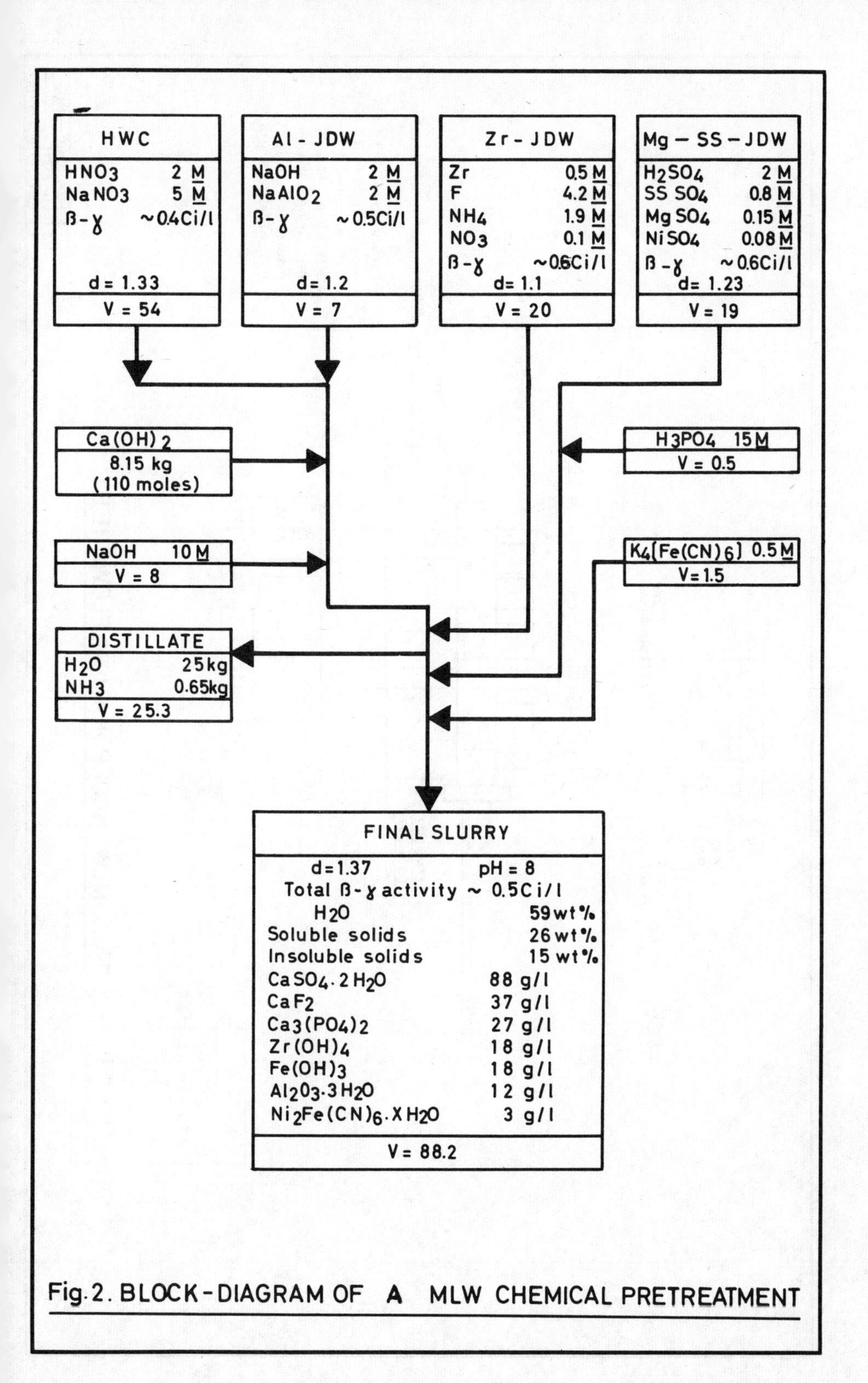

Fig. 2. BLOCK - DIAGRAM OF A MLW CHEMICAL PRETREATMENT

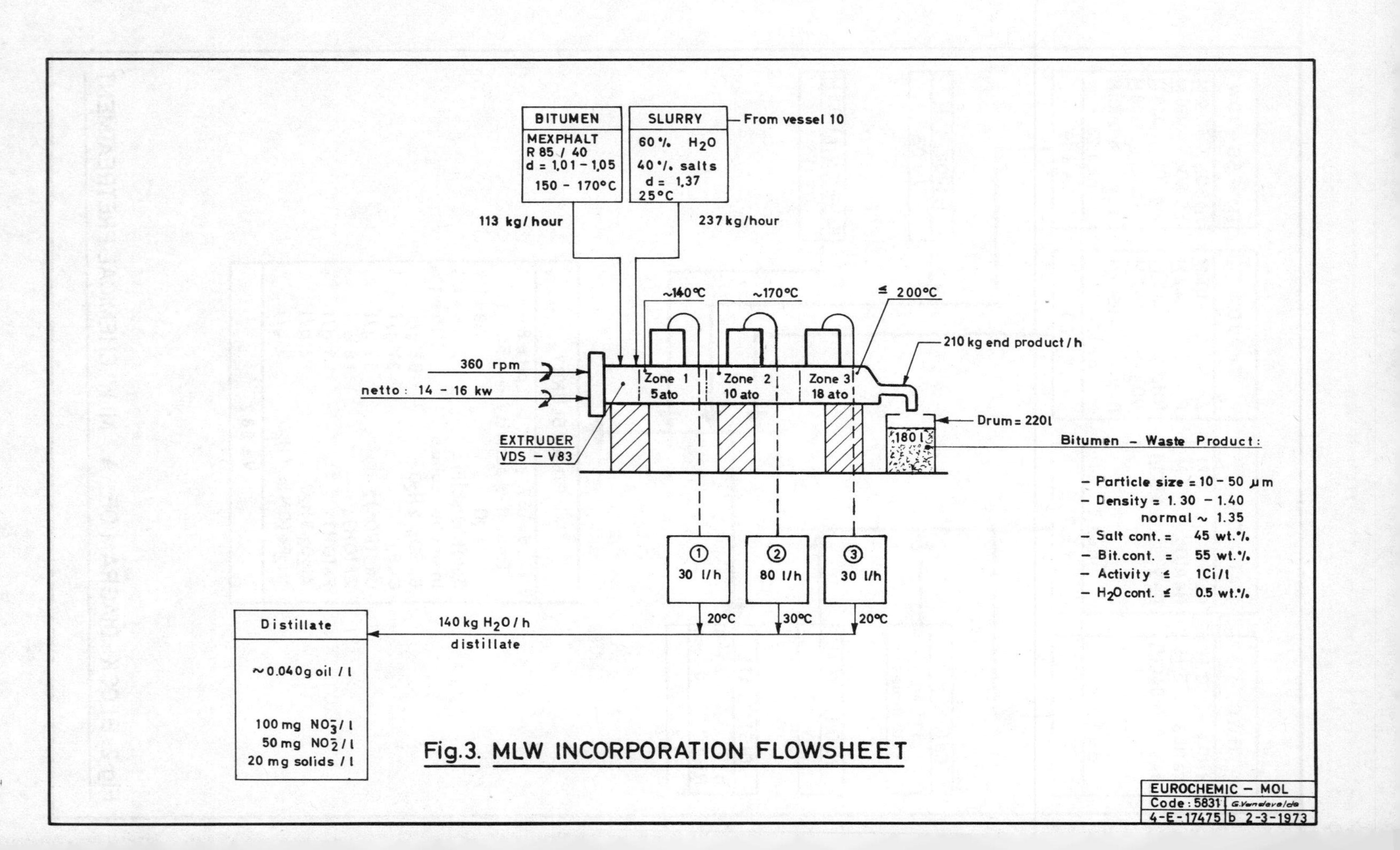

Fig.3. MLW INCORPORATION FLOWSHEET

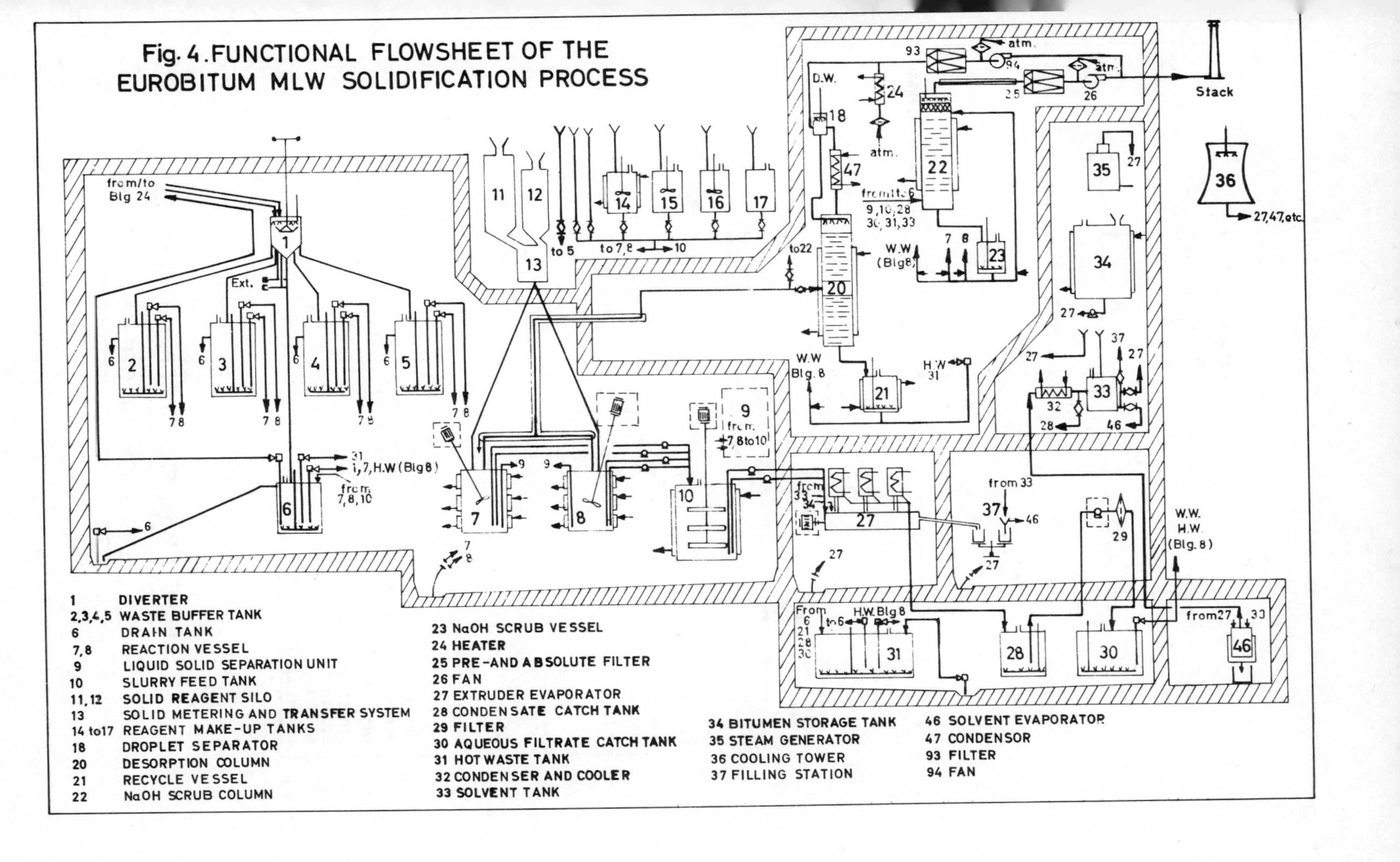

Fig. 4. FUNCTIONAL FLOWSHEET OF THE EUROBITUM MLW SOLIDIFICATION PROCESS

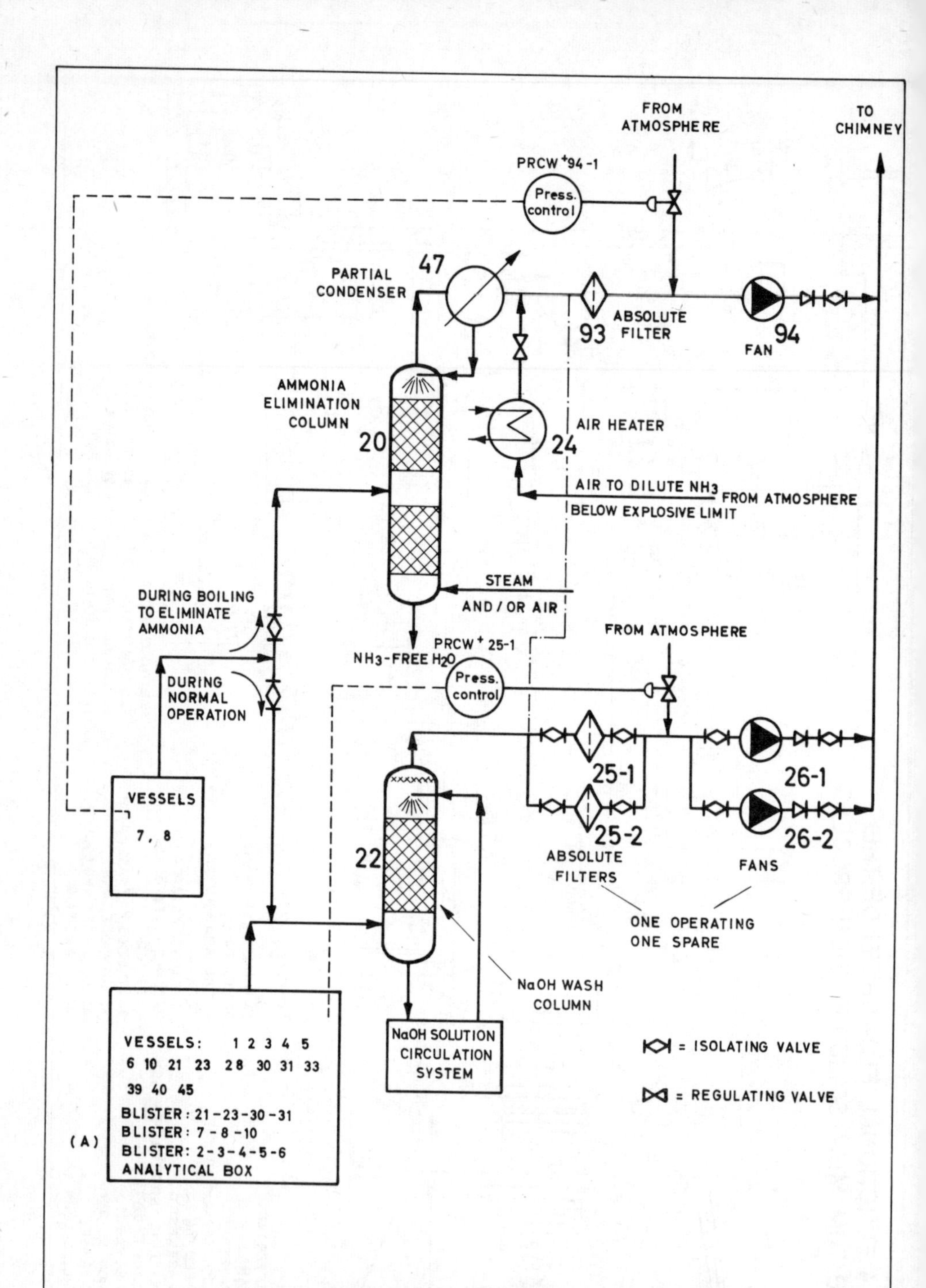

Fig.5. SIMPLIFIED DIAGRAM OF VESSEL VENTILATION AND OFF-GAS SYSTEM

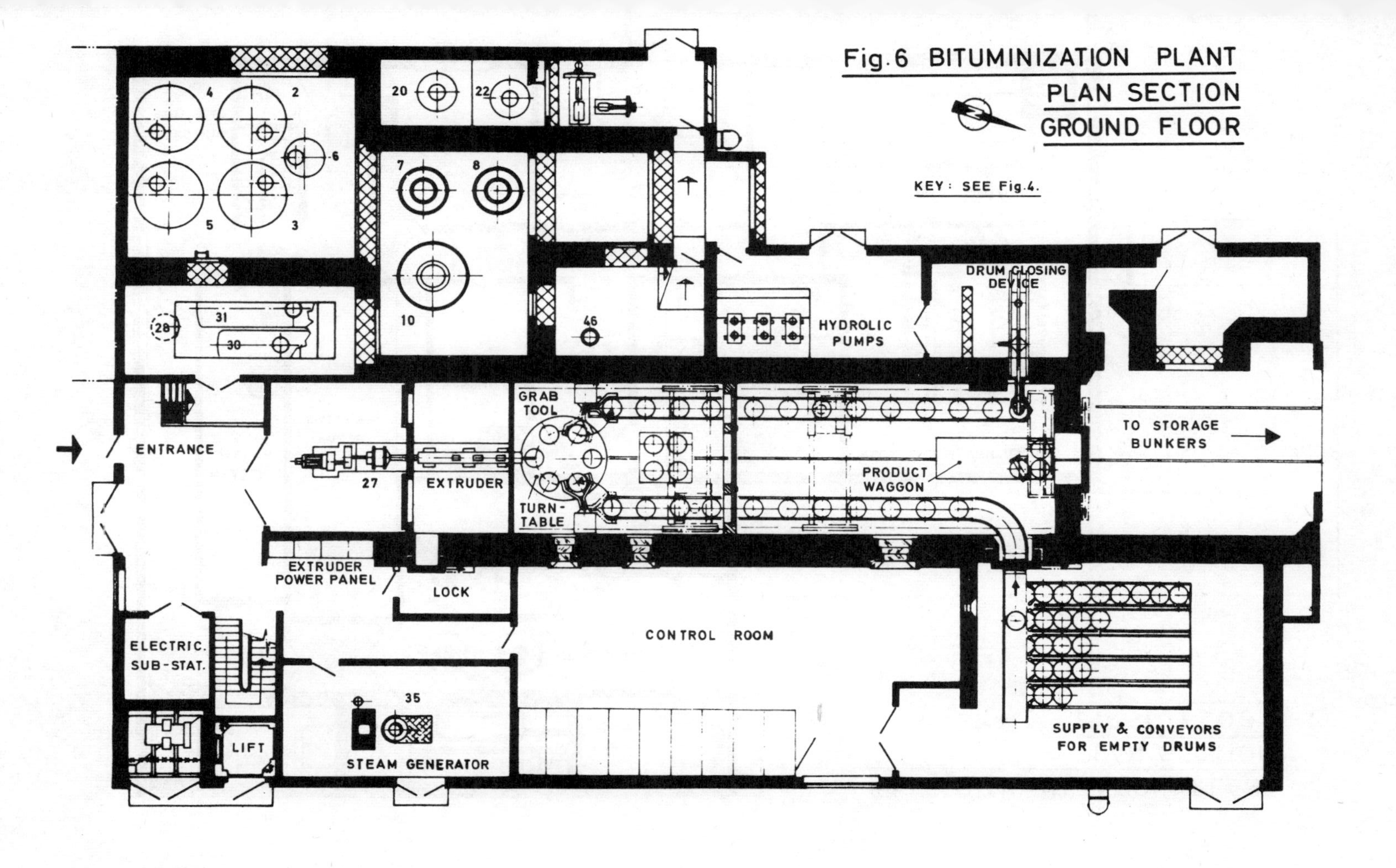

Fig. 6 BITUMINIZATION PLANT PLAN SECTION GROUND FLOOR

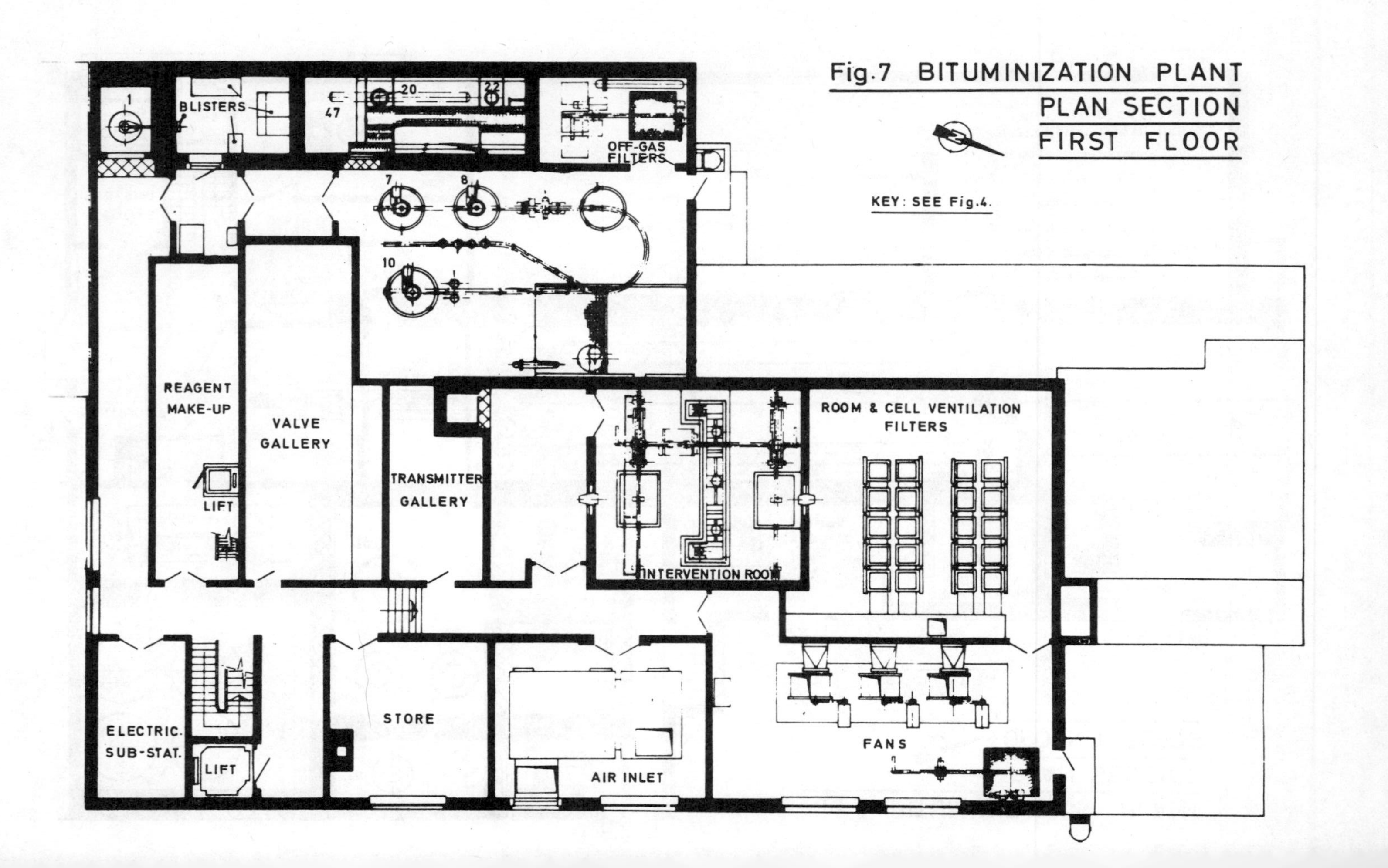

Fig.7 BITUMINIZATION PLANT
PLAN SECTION
FIRST FLOOR

Discussion

P.W. KNUTTI, Switzerland

What has been foreseen to prevent possible fire ?

H. ESCHRICH, Eurochemic

At Eurochemic the following measures have been taken to prevent fire risks during the performance of the bituminisation process :

- the waste solutions to be solidified are periodically analysed - though their composition most probably remains unchanged during the coming years - to detect and to determine materials which could directly or indirectly (e.g. by catalysis) contribute to a fire risk ;

- the slurries are analysed to verify, among other things, that the pH of the mother liquor has the foreseen value of between 8 and 9 to prevent rapid saponification or other decomposition reactions of organic compounds ;

- using the slurry, prepared according to a certain flowsheet, to produce a bitumen-waste product sample and subjecting it to a differential thermo-analysis to detect whether strong exothermic reactions have to be expected at the incorporation temperatures ;

- the extruder is heated by steam of a max. pressure of 20 atm. which prevents any local overheating of the contents within the module ;

- moreover a strong ventilation (suction) above the opening of the product drum, beneath the product outlet tube and above the product collection funnel avoids the accumulation of inflammable or explosive gas mixtures ;

- a water injection device permits cooling down the bitumen-waste product in case an abnormal evolution of gases in the product drums is observed ;

- in all rooms and cells where burnable materials are present or handled fire detection and fire fighting equipment have been installed.

G. ENGELHARDT, F.R. of Germany

You accumulated 2200 m^3 MLW during the operation time of Eurochemic :

a) Is that all the waste you got during that time from the beginning of operation ?

b) What is your design input for the bitumen plant ?

c) What did you calculate within this input-volume (MLW) for decontamination waste coming from interventions in your bitumen-plant ?

H. ESCHRICH, Eurochemic

a) During the entire fuel reprocessing period Eurochemic has produced 945 m^3 of decladding solutions and a little more than 1200 m^3 of hot waste concentrate. Furthermore a part of the hot waste concentrate has recently been concentrated so that the MLW volume of 2200 m^3 at the end of 1974 has been reduced to about 2100 m^3 in the beginning of May 1976. This is, indeed, all the liquid MLW produced at Eurochemic.

b) Based on 45 working weeks per year and a 60 % availability of the full operation capacity of the extruder the bituminization plant can treat 650 m^3 of MLW solutions per year in the three-shift schedule or 950 m^3 per year in a four-shift schedule.

c) During reprocessing operations the hot waste concentrate included the decontamination wastes. The average production rate of hot waste concentrate solutions amounted to 15 m^3/month.

Compared to the decontamination wastes generated in the reprocessing plant the amounts produced in the bituminization installations during the solidification period are expected to be much smaller and have therefore not been taken into consideration.

G. ENGELHARDT, F.R. of Germany

With respect to the aim to run or operate the extruder under "steady state conditions", what is the volume of your slurry tank ?

Could you also give some information about the operation schedule of this tank. Do you exercise discontinuous operation of the slurry tank, that means, at first filling then later conditioning, and finally feeding the extruder ?

H. ESCHRICH, Eurochemic

As described in the paper the slurry tank from which the extruder is fed has a volume of 7 m^3 allowing 1.5 days of continuous operation.

Besides the slurry feed tank two identical slurry preparation vessels of 3 m^3 each are installed from which 2.5 m^3 slurry batches could be transferred every five hours, if required. Thus, the large capacity of the slurry feed tank allows continuous operation of the extruder whilst the two slurry preparation vessels (reaction vessels for chemical pretreatments) are operated batchwise.

After each addition of a slurry batch to the slurry feed tank, whose contents are homogenized by a powerful mechanical stirrer, a sample is taken and analyzed to ensure proper operation conditions.

N. FERNANDEZ, France

Mr. Chairman, following this paper on such a remarkable installation, it is necessary to ask :

- what is the cost of the installation,

- what is the expected operating cost ?

E. DETILLEUX, Eurochemic

1. Treatment installation (Eurobitume) :

   - basic investments - BF ~ 200 M ($ 5.10^6)

   - modifications, cold tests - BF ~ 15 M ($ $0.37.10^6$)

   - operation - BF ~ 100-120,000/m^3 of treated effluents excluding investments.

2. Storage :

   - basic investments BF 130 M ($ 3.25 M), including
     - the liaison corridor for 6 bunkers ;
     - the reception station for waste produced outside the bituminisation plant ;
     - handling equipment ;
     - two bunkers (10,000-12,000 drums)

   - additional bunker, including preparation of the site ~ BF 25 M ($ $0.625.10^6$, price 1976).

J. STORRER, Belgium

I should like to add a comment concerning present costs. BelgoNucléaire has just signed a contrat with PNC Japan for an installation which is similar to Eurobitume and Eurostorage, for Tokai Mura. Its price is of the order of $ 20 M.

H. DWORSCHAK, Italy

What will be the concentration of Pu in the solid ?

H. ESCHRICH, Eurochemic

If all types of wastes were treated by applying a flowsheet involving a proportional mixing of the waste solutions presently stored, then the plutonium concentration in the final product would be about 1 mg/l product.

H. DWORSCHAK, Italy

What concentration of Pu could be considered reasonable for this kind of solid ?

H. ESCHRICH, Eurochemic

The presence of various alpha-emitting plutonium isotopes or other alpha-emitters should not cause radiation damage endangering the integrity of the final product to an unacceptable degree. The resistance of various types of bitumen and bitumen-salt mixtures towards alpha radiation is presently the subject of investigations being performed by GfK, Karlsruhe, in co-operation with Eurochemic to evaluate the maximum permissible specific alpha activity of a wide variety of bitumen-waste products ; this maximum value is presently unknown, but, can be expected to be several times higher than the alpha activity in Eurochemic's products. Ideally the bitumen wastes should not contain any appreciable amounts of alpha emitters so that a "short-lived" waste could be obtained which need not be classified and disposed of as alpha-waste.

E.P. UERPMANN, F.R. of Germany

What can you do, if you get problems of overflowing bitumen caused by increasing the volume of radiolysis products ?

H. ESCHRICH, Eurochemic

The 220 l drums are filled with approximately 180 l bitumen waste product only, to reserve about 20 % of the available space for an eventual volume increase of the drum contents. External irradiation of simulated Eurochemic products - prepared in the bituminization facility at Marcoule - by a cesium 137 source of 2.6×10^6 rad/h up to an integrated dose of 8×10^9 rad showed a maximum volume increase of 1 %. As the total irradiation dose the products of max. 1 Ci/l will receive is lower than 10^9 rad we do not expect any problems due to expansion of the products. In this connection it might be of interest to mention that our product drums are closed by a special cover, which is not gas-tight.

R. SIMON, CEC

What means of leak detection and maintenance intervention are provided in the shielded solidification cells ?

H. ESCHRICH, Eurochemic

In the case of an accidental liquid leak, drip-trays have been installed which are provided with a detection system for liquids coupled to an alarm system.

The drip-trays are made of stainless steel and have a volume at least equivalent to the volume of the largest vessel installed in the respective cell. All drip-trays can be decontaminated and the liquids can be removed by means of steam-jets or pumps.

In the solidification cells in which the active bitumen-salt mixtures are handled, certain interventions can be carried out remotely by means of the telemanipulator.

Radioactive materials have to be removed from the cells to the required extent before any direct maintenance can be carried out.

Sufficient connections have been installed for decontamination and cleaning purposes and throughlets have been provided for the introduction of special cleaning equipment.

R. SIMON, CEC

What radiation levels are expected in the Eurostorage service corridor when the storage bunkers are full ?

H. ESCHRICH, Eurochemic

The exposure rate in the service corridor has been calculated to be 0.25 mrem/h on the assumption that the bunkers are filled with 5000 waste product drums.

Discussion

P.W. KNUTTI, Suisse

Qu'a-t-on prévu pour empêcher qu'un incendie ne se produise ?

H. ESCHRICH, Eurochemic

A l'usine d'Eurochemic, les mesures suivantes sont prises contre les risques d'incendie pendant le déroulement du procédé de bitumage :

- les solutions de déchets à solidifier sont analysés périodiquement - bien que, selon toute vraisemblance, leur composition ne doive pas se modifier au cours des années à venir - en vue de déceler et de déterminer les matières qui pourraient, directement ou indirectement (par catalyse, notamment), contribuer à un risque d'incendie ;

- les boues sont analysées en vue de vérifier, notamment, que le pH de la solution-mère a bien la valeur prévue (entre 8 et 9) pour empêcher une saponification rapide ou d'autres réactions de décomposition des composés organiques ;

- on utilise les boues, obtenues selon un certain diagramme, pour produire un échantillon du déchet incorporé dans du bitume et on le soumet à une thermo-analyse différentielle afin de déterminer s'il faut s'attendre à de fortes réactions exothermiques aux températures d'incorporation ;

- l'extrudeuse est chauffée par la vapeur à une pression maximale de 20 atmosphères, ce qui empêche toute surchauffe locale du contenu à l'intérieur du module ;

- en outre, une forte ventilation (aspiration) au-dessus de l'ouverture du fût de produits, qui se trouve en dessous du tube de sortie des produits et en dessus de l'entonnoir destiné à recueillir les produits, évite l'accumulation de mélanges de gaz inflammables ou explosifs ;

- un dispositif d'injection d'eau permet de refroidir le déchet incorporé dans du bitume au cas où l'on observerait une évolution anormale des gaz dans les fûts de produits ;

- toutes les salles et cellules dans lesquelles des matières combustibles sont présentes ou manipulées ont été dotées d'appareils de détection des incendies et de lutte contre l'incendie.

G. ENGELHARDT, R.F. d'Allemagne

Le volume de déchets de moyenne activité accumulés pendant la durée d'exploitation de l'usine d'Eurochemic a été de 2200 m^3 :

a) S'agit-il de la quantité totale de déchets produits depuis la mise en service de l'usine ?

b) Quel est le volume d'entrée prévu pour l'installation de bitumage ?

c) Quelle fraction de ce volume d'entrée (déchets de moyenne activité) avez-vous attribué, dans vos calculs, aux déchets provenant des interventions effectuées dans l'installation de bitumage ?

H. ESCHRICH, Eurochemic

a) Pendant toute la période de retraitement du combustible, l'usine d'Eurochemic a produit 945 m^3 de solutions de dégainage et un peu plus de 1200 m^3 de concentrats de déchets chauds. Une partie des concentrats de déchets chauds vient d'être soumise à une nouvelle opération de concentration, de sorte que le volume de déchets de moyenne activité, qui représentait 2200 m^3 à la fin de 1974, a été ramené à 2100 m^3 environ au début du mois de mai 1976. Il s'agit là, en fait, de la totalité des déchets liquides de moyenne activité produits à Eurochemic.

b) Compte tenu de quarante-cinq semaines de travail par an et dans l'hypothèse où l'extrudeuse pourrait être utilisée au maximum de sa capacité pendant 60 % du temps, l'installation de bitumage peut traiter 650 m^3 de solutions de déchets de moyenne activité par an, si l'on prévoit un travail à trois postes, ou, 950 m^3 par an, si l'on prévoit un travail à quatre postes.

c) Pendant les opérations de retraitement, les effluents de décontamination étaient compris dans les concentrats de déchets chauds. Le taux moyen de production de solutions de concentrats de déchets chauds représentait 15 m^3 par mois.

Par rapport au volume d'effluents de décontamination enregistré dans l'installation de retraitement, les quantités d'effluents de ce type produites dans les installations de bitumage au cours de la période de solidification seront vraisemblablement beaucoup moins élevées, aussi n'ont-elles pas été prises en considération.

G. ENGELHARDT, R.F. d'Allemagne

L'objectif visé étant de faire fonctionner l'extrudeuse en régime stationnaire, il y a lieu de savoir quel est le volume de votre réservoir d'alimentation en boues.

D'autre part pourriez-vous nous donner quelques renseignements au sujet du programme d'exploitation de ce réservoir ? L'utilisez-vous de façon discontinue ? En d'autres termes, procédez-vous d'abord au remplissage, puis au conditionnement et enfin à l'alimentation de l'extrudeuse ?

H. ESCHRICH, Eurochemic

Comme il a été indiqué dans la communication, le réservoir servant à alimenter l'extrudeuse en boues a un volume de 7 m^3, ce qui permet de l'utiliser en continu pendant un jour et demi.

A côté du réservoir d'alimentation en boues, on a installé deux cuves identiques de 3 m^3 chacune pour la préparation des boues, à partir desquelles il serait possible, le cas échéant, de transférer des "lots" de boues de 2,5 m^3 toutes les cinq heures. Dans ces conditions, la capacité élevée du réservoir d'alimentation en boues permet d'utiliser l'extrudeuse en continu, alors que les deux cuves de préparation des boues (cuves de réaction pour les prétraitements chimiques) sont utilisées de façon intermittente.

Chaque fois qu'un "lot" de boues a été introduit dans le réservoir d'alimentation en boues, dont le contenu est homogénéisé à l'aide d'un puissant agitateur mécanique, un échantillon est prélevé et analysé afin de s'assurer des bonnes conditions de fonctionnement.

N. FERNANDEZ, France

M. le Président, à la suite de cet exposé sur une installation remarquable, une question s'impose :

- Quel est le coût de l'installation ?
- Quel est le coût d'exploitation prévu ?

E. DETILLEUX, Eurochemic

1. Installation de traitement (Eurobitume)
   - Investissements de base : F.B. ~ 200 M ($ 5.10^6)
   - Modifications, essais à froid : F.B. ~ 15 M ($0,37.10^6$ $)
   - Exploitation : ~ 100 à 120.000 F.B./m^3 d'effluents traités hors investissement.

2. Stockage - Investissements de base : F.B. 130 M ($ 3,25 M) comprenant :
   - le couloir de liaison pour six "bunkers"
   - la station de réception des déchets produits en dehors de l'installation de bitumage
   - toute la manutention
   - deux bunkers (10.000 à 12.000 fûts)
   - Bunker additionnel y compris le chantier à ouvrir ~ 25 M F.B. ($ $0,625.10^6$ prix 1976).

J. STORRER, Belgique

J'aimerais ajouter une précision concernant les prix actuels. BelgoNucléaire vient de conclure un contrat avec PNC (Japon) pour une installation similaire à EUROBITUM + EUROSTORAGE pour Tokai Mura ; son prix est de l'ordre de 20 millions de $.

H. DWORSCHAK, Italie

Quelle sera la concentration de plutonium dans le solide ?

H. ESCHRICH, Eurochemic

Si tous les types de déchets étaient traités conformément à un diagramme prévoyant un mélange proportionnel des solutions de déchets actuellement stockées, la concentration de plutonium dans le produit final serait d'environ 1 mg/l de produit.

H. DWORSCHAK, Italie

Quelle est la concentration de plutonium qui pourrait être considérée comme raisonnable pour ce type de solide ?

H. ESCHRICH, Eurochemic

La présence de divers isotopes de plutonium émetteurs alpha ou d'autres émetteurs alpha ne devrait pas provoquer de dommages mettant en danger de façon inacceptable l'intégrité du produit final. La résistance des divers types de bitumes et des mélanges de sels et de bitume au rayonnement alpha fait présentement l'objet de recherches effectuées par GfK à Karlsruhe, avec la collaboration d'Eurochemic, en vue d'évaluer l'activité spécifique alpha maximale admissible de toute une gamme de déchets incorporés dans du bitume ; cette valeur maximale n'est pas connue à l'heure actuelle mais elle sera vraisemblablement plusieurs fois supérieure à l'activité alpha contenue dans les produits d'Eurochemic. Au mieux, les déchets incorporés dans du bitume ne devraient pas renfermer de quantités importantes d'émetteurs alpha, de façon à ce qu'il soit possible d'obtenir un déchet à courte période qui n'ait pas besoin d'être classé dans la catégorie des déchets alpha et évacué comme tel.

E.P. UERPMANN, R.F. d'Allemagne

Si l'augmentation du volume des produits de radiolyse fait déborder le bitume, que pouvez-vous faire pour surmonter les difficultés qui en résultent ?

H. ESCHRICH, Eurochemic

Les fûts de 220 l ne sont remplis de déchets incorporés dans du bitume qu'à raison de 180 l environ par fût, afin de réserver de l'ordre de 20 % de l'espace disponible au cas où le contenu des fûts viendrait à augmenter de volume. Une expérience d'irradiation externe portant sur des produits du type Eurochemic (qui avaient été préparés dans l'installation de bitumage de Marcoule), au cours de laquelle on a utilisé une source de césium 137 de 2,6 x 10^6 rads/h de manière à atteindre une dose cumulée de 8 x 10^9 rads, a montré que l'augmentation de volume ne dépassait pas 1 %. Comme la dose totale d'irradiation à laquelle seront soumis les produits de 1 Ci/l au maximum est inférieure à 10^9 rads, nous ne prévoyons pas de problèmes dus à une dilatation des produits. A cet égard, il pourrait être intéressant de signaler que nos fûts de produits sont fermés par un couvercle spécial, qui n'est toutefois pas étanche aux gaz.

R. SIMON, CCE

Quels moyens ont été prévus pour détecter les fuites et procéder aux opérations d'entretien dans les cellules de solidification blindées ?

H. ESCHRICH, Eurochemic

En cas de fuite accidentelle de liquide, on a installé des lèchefrites qui sont munis d'un système de détection des fuites de liquide, raccordé à un dispositif d'alarme.

Les lèchefrites sont en acier inoxydable et leur volume est au moins équivalent à celui de la plus grande cuve installée dans la cellule correspondante. Toutes les lèchefrites peuvent être décontaminées et les liquides peuvent être éliminés au moyen de jets de vapeur ou de pompes.

Dans les cellules de solidification où sont manipulés les mélanges radioactifs de sels et de bitume, certaines opérations peuvent être effectuées à distance au moyen du télémanipulateur.

Les matières radioactives doivent être éliminées des cellules dans toute la mesure nécessaire avant qu'il soit possible d'effectuer une opération directe d'entretien.

On a installé suffisamment de raccordements à des fins de décontamination et de nettoyage, et des orifices de passage ont été prévus pour l'introduction d'appareils spéciaux de nettoyage.

R. SIMON, CCE

Quels niveaux de rayonnement prévoyez-vous dans le couloir de liaison EUROSTORAGE lorsque les enceintes de stockage sont pleines ?

H. ESCHRICH, Eurochemic

Le taux d'exposition dans le couloir de liaison a été évalué à 0,25 mrem/h dans l'hypothèse où les enceintes contiendraient 5.000 fûts de déchets.

COMMISSIONING AND START-UP TESTS OF EUROCHEMIC'S WASTE BITUMINIZATION FACILITY

M. Demonie

Eurochemic

Mol- Belgium

ABSTRACT

This paper describes the commissioning tests of the Eurobitum Facility performed during the period October 1974 - December 1975 and the start-up experience gained during the period December 1975 - April 1976.

RESUME

Ce document est une description des essais de réception de l'unité Eurobitume pour la période Octobre 1974 - Décembre 1975 et le compte-rendu des opérations de démarrage à froid effectuées au cours de la période Décembre 1975 - Avril 1976.

1. INTRODUCTION

The commissioning and start-up tests of the Eurochemic Waste Bituminization Facility started in November 1974 by the usual rinsing operations followed by the pressure and tightness tests of the equipment and its calibration.

This first phase has been followed by a series of functional tests aiming to check the basic performances of the most important subunits.

This phase, which is still going on, will be followed by a "cold test" phase aiming to test the entire installation operated such as it will be during routine operation.

The present paper summarizes the main informations collected during the programme, some of them having led to some modifications of the originally installed equipment.

2. RINSING, PRESSURE AND TIGHTNESS TESTS

Rinsing has been performed by means of filtered water, demineralized water or compressed air, according the nature of the circuits.

Hydrostatic and pneumatic pressure tests applied pressure 1.5 times the design pressure during one hour.

Leaktests were performed at 0.25 kg/cm^2 air pressure, soap-bubble being used to detect eventual leaks.

No special problems have been encountered during those tests.

3. CALIBRATION OF VESSELS, JETS, AIR-LIFTS AND SAMPLING SYSTEMS

Vessels were calibrated with water according to the Eurochemic specifications for process vessels "Class C".

Jets have been calibrated with water at a steam pressure of 8 ATA.

The calibration of the air-lifts led to the modification of two air-lifts only.

A sampling system is installed in 12 vessels; the tests have shown that the slurry sampling system had to be modified in order to improve the homogeneity of the slurry collected in the sample pot. Fig. 1 shows the sampling system originally installed, and Fig. 2 the system as modified. As it can be seen, the air-slurry mixture will now be sucked through the sample pot, prior to reach the air-lift pot where it will be separated.

4. VESSEL VENTILATION

Tests have indicated cross connections between the ventilation loops to be used on one hand when ammonia desorption is performed, on the other hand when no ammonia removal is required.

Some other improvements have been required to secure the satisfactory underpressure and air flow.

In Fig. 3 is represented the normal vessel ventilation system while Fig. 4 represents the ammonia vessel ventilation system.

5. AMMONIA ELIMINATION TESTS

5.1 Introduction

Ammonium ions are present in considerable amounts in the Zirflex solution and the hot waste concentrate (HWC).

To avoid a possible uncontrollable exothermic reaction of ammonium nitrate with bitumen and other organic matters present in the MLW, the ammonium is eliminated in form of ammonia by boiling and agitating the alkaline slurry in the reaction vessel 7 or 8.

The water vapour and ammonia are separated in a heated packed column 20 connected to a reflux cooler 47 (see Fig. 4).

The progress of the ammonium removal can be controlled by measuring the conductivity of the distillate flowing back from the cooler 47 to the column 20.

The ammonia released in the reaction vessel can be treated in two ways :

- by desorption and release to the stack,
- by absorption into water and transfer to building 8.

5.2 Performed modifications

The following modifications were necessary to assure a correct performance of the system :

- addition of Δp measurements over column 20, condensor 47, droplet separator 18,
- addition of flow measurement of the recycled distillate,
- installation of a distillate recirculation air-lift from condensor 47 to column 20,
- variable demiwater supply (FCV 47-3) to column 20, to guarantee under all circumstances a sufficient scrubflow.

5.3 Ammonia desorption tests

The most important factor to determine is obviously the ammonium remaining in the solution of the reaction vessel at the end of the desorption operation.

It is also important that the ammonium content in the raffinate in the vessel 21 is as low as possible which will prove the performance of the column 20.

The process conditions aimed at during the series of tests were :

- evaporation rate ca. 300 l/h,
- heating of jacket on column 20, and on the off-gas line from the reaction vessel to column 20,
- condensor outlet temperature set at 45°C,
- 75 mm underpressure in the reaction vessel,
- recycled flow to column 20 kept at least at 300 l/h, adjusted by means of demiwater (FCV 47-3),
- cooling of jacket on vessel 21,
- prefilter and absolute filter in filtercasing 93.

The tests have demonstrated the high efficiency of the ammonia elimination system. Only ca. 0.1 % of the ammonium remains in the solution of the reaction vessel at end of the desorption test, while about 2 % of the total ammonium is retained in the raffinate.
The maximum permissible ammonium concentration remaining in the slurry solution will be determined by differential thermoanalyses experiments and this will determine the necessity of prolonged boiling or extra sparging in the reaction vessel.
The ammonium remaining in the solution will probably vary depending on the physical characteristics of the mixture.
This effect will be examined in more detail during the cold tests.

5.4 Ammonia absorption tests

The advantage of this technique is that the possible explosion hazard in the off-gas circuit due to the ammonia-air mixture will be practically eliminated.

The disadvantages however are :

- risk of clogging the off-gas filters due to ammonium nitrate deposition on these filters,

- the high salt content in the liquid warm waste.

The process conditions described above (§ 5.3) have been modified as follows :

- no heating of off-gas line to column 20,

- cooling of jacket on column 20,

- maximum cooling on the condensor 47,

- demiwater flow to column 20 kept at 150 l/h (1st test) and 200 l/h (2nd test).

The tests have demonstrated a poor retaining efficiency of the system. In both tests only ca. 50 % of the total ammonia is retained in the raffinate, while the rest leaves the system to the stack. Consequently it is not considered any longer.

5.5 Air dilution system

The uncondensed off-gas leaves the condensor 47 at a temperature of ca. 45°C. The off-gas will contain a maximum of 49 kg/h or 63 Nm^3/h of ammonia when treating pure Zirflex solution. This off-gas must be diluted with a minimum of 600 Nm^3/h of preheated air (45-80°C) before its passage through filter 93 and its subsequent release to the stack.

The air dilution system guarantees the safe operation i.e. :

- the ammonia concentration in the air is kept below the explosion limit of 14 vol.% in air,

- the temperature of the ammonia air mixture is kept above 40°C (at 40°C the "explosive" range of an ammonia air mixture saturated with water vapour is very narrow = 20 - 25 vol. % of ammonia in air).

In order to guarantee a sufficient air dilution (600 Nm^3/h), the installed heat exchanger 24 had to be changed for another with a lower pressure drop, resulting in a higher dilution flow for the same underpressure at the inlet.

From the tests it can be deduced that :

- there is a linear relation between the temperature and the flow of the dilution air, at 900Nm^3/h the temperature is still 52°C, at 600Nm^3/h the temperature is 65°C,
- the air dilution flow is a function of the evaporation rate, at 300 l/h evaporation rate the air dilution flow is 840 Nm^3/h, at 100 l/h evaporation rate the air dilution flow is still 660 Nm^3/h.

The tests were made keeping an underpressure of 75 mm WG in the vessel 7.

5.6 Conductivity measurement in the recycled distillate

The object of the conductivity recorder is to follow the course of the ammonia removal and to indicate the end of the reaction.

The relation between the conductivity recorder reading in the distillate and the ammonium concentration in the boiling solution is presented in Fig. 5.

Since the conductivity calibration curve of ammonium hydroxide (Fig. 6) has a maximum at 3 M, there should be two maxima on the recording of the conductivity during the evaporation i.e. at the beginning when the distillate overpasses the concentration of 3 M NH_4OH and near the end when the concentration in the distillate decreases under 3 M NH_4OH.

In reality three peaks were recorded, the two first peaks due to the here above mentioned evolution of the ammonium concentration in the distillate, the last one due to the reconcentration of ammonium at the top of the column 20 when there is no evaporation anymore in the reaction vessel and consequently the dilution with steam from the reaction vessel ceases.

6. DRUM HANDLING SYSTEM TESTS

6.1 General

The Bitumen Incorporated Product (BIP) is collected in "chromized" steel drums of 220 l nominal capacity which are filled to 80 % in order to allow some swelling, if any.

After cooling, the drums are closed with a cover-lid which is not fully tight in order to allow the escape of any gases produced by radiolysis.

The drums are transported by means of a conveying system represented in Fig. 7. The cooling down speed of the waste product can be deduced from Fig. 8, representing the temperature of the product in function of time after filling.

In room 015, five rows of 8 empty drums are installed for feeding the conveyor 62. The empty drums are transported by a roller conveyor from room 015 to cell 005. A hydraulic grab-tool 97 transfers the drums one by one to the turning table, where the drums receive the bitumen waste product.
The filled drums are transferred by the hydraulic grab-tool 98 from the turning table 37 to a bandplate conveyor 64. On conveyor 64 the possibility to sample the "BIP" has been provided. The conveyor 64 transports the full drums to its end position, from where a third hydraulic grab-tool 95 places the drums one by one on a waggon. Prior to be put on the waggon a coverlid is placed on each drum.

6.2 Performed tests and results

The drum handling system can be operated manually or automatic. The tests were made with empty drums and with loaded drums (400 kg max.) as well in manual as in automatic.
The drum closing device was tested simultaneously.

Due to the fact that the sampling position of the "BIP" and the drum closing device position were modified compared to the original design, the step on the blade conveyors 64 and 65 had to be adapted accordingly.

The major difficulties were encountered with the instrumentation of the drum handling system.

After the necessary repairs, the system was tested during five consecutive days in full automatic operation, the drums being filled with water on the blade conveyors 64 and 65.

6.3 Transport of the drums by waggon to the bunker

The drums are transported by waggon from cell 013 to the storage facility. The waggon has a total capacity of twelve drums.

7. SOLID REAGENTS SILOS AND METERING UNIT TESTS

Calcium hydroxide and barium hydroxide are the two solid reagents used during the chemical pretreatment process.

Calcium hydroxide is delivered to the site by truck from which it is transferred pneumatically to its storage silo (52 m^3 or 30 t capacity). From the storage silo, it is transferred pneumatically via a cyclone to the weighing bin.

Barium hydroxide is received in bags. It is transferred pneumatically to its weighing bin, passing on its way respectively a cyclone and a grinder.

The powder in the weighing bins is kept fluidized by means of a fan, allowing smooth transfer by gravity to the reaction vessels 7 or 8.

A functional test of the system has been made with only minor difficulties which have been overcome.

8. BITUMINIZATION UNIT TESTS

The slurry is metered from feed vessel 10 to the extruder, where it is mixed with hot bitumen.

8.1 The extruder

8.1.1 General

A 4 screws extruder (manufactured by Werner and Pfleiderer, Stuttgart) has been selected for the incorporation of the slurry into the bitumen.

The distillate leaving the condensors is filtered to remove the entrained oil.

The incorporation temperature at the extruder outlet is kept between 180 - 200°C and this for two reasons :

a) maximum 200°C in order to exclude any risk of uncontrollable exothermic reaction between the nitrates and the bitumen,

b) minimum 180°C in order to reach a low final water content in the product.

8.1.2 Test runs performed and conclusions

A series of tests have been performed to examine the different important parameters.

a) Some functional tests were made first with hot bitumen only. The aim was to check the good functioning of the different instruments of the extruder and its general performance.

b) A first series of tests were made afterwards with the following components :

Slurry : sodium nitrate solution at 40 wt-%, in some cases with some wt-% of $NaNO_2$.

Bitumen : Mexphalt 10-20.

Different ratios of solids/bitumen were tried out at different flowrates and temperatures. All incoming and outgoing streams were sampled. The distillate was analysed for pH, $NaNO_3$, $NaNO_2$ conductivity and oil content. The "BIP" was analysed for wt-% $NaNO_3$, wt-% $NaNO_2$, wt-% bitumen, density and H_2O content.

A summary of the results obtained is given in the table hereafter :

Properties & compositions of the distillate		1st dome	2nd dome	3rd dome
pH		4-9	3-7	2-5.5
Conductivity	mS/cm	0.6-2	1-2	1-7.5
$NaNO_3$	g/l	0.3-1.6	0.1-0.9	0.03-0.3
$NaNO_2$	mg/l	0.6-50	1-20	2-20
Decontamination Factor		400-2000	$600-6.10^3$	$2.10^3-2.10^4$
Oil content	wt-%	0.001	0.002-0.013	0.003-0.02
Distillate	V %	49	31.7	19.3
t° region	°C	150	175	175

Properties & compositions of the end product		Slurry composition $NaNO_3$ 40 wt-% $NaNO_2$ nihil	Slurry composition $NaNO_3$ 38.5 wt-% $NaNO_2$ 0.60 wt-%
$NaNO_3$	wt-%	15 - 75	
$NaNO_2$	wt-%	20 - 60 ppm	0.14 - 0.9 wt-%
Mexphalt 10-20	wt-%	25 - 85	
Density at 20°C	g/ml	1.15 - 1.55	
H_2O	wt %	0.07 - 3.5	

The water content of the BIP is depending very much on the temperature in the last section of the extruder. At a temperature of 180°C the water content is around 2 wt-%. The high wt-% $NaNO_3$ (75 wt-%) results from an irregular slurry feed flow. Due to this high salt composition in the end-product the outlet of the extruder was plugged. The decontamination factor in the distillate is expressed as the ratio of the concentrations of $NaNO_3$ in the slurry and in the distillate.

8.1.3 Encountered problems

- Heating system

To attain the required temperature profile in the extruder at 140 kg/h evaporation rate, the Velan steam traps of ½" size proved to be undersized. Moreover, their location above the extruder is not in favour of a good performance.

- Feed inlet part of the extruder

On several occasions the slurry feed line plugged with bitumen. Due to a varying pressure in the feed inlet part, the flows of the bitumen and the slurry are fluctuating.

The original cover-lid was removed and replaced by a flange type cover-lid, provided with a ventilation line.

Entrainment of bitumen at a rate of 2-15 l/h via the ventilation line was noticed, depending on the slurry feed temperature (20-80°C).
To suppress this entrainment of bitumen, the temperature profile on the feed inlet part and the 1st section of the extruder was lowered in order to minimize the evaporation in that part.

- Leak on stuffing boxes of the extruder screws

The origin of this leak was two-fold :

a) aged packing material (3 years between supply and 1st operation,

b) pressure build-up in the feed inlet part.

The technique to renew the packing material is rather complicated, a redesign of the stuffing box should be useful.

- Plugging of the bitumen supply line

The bitumen supply line plugged due to insufficient heating on the last part of the line : the heating jacket had to be prolonged.

- Plugging of the extruder outlet

The reason of this plugging was the too high salt content in the bitumen incorporated product (75 wt-% salts).

8.2 Bitumen storage

The molten bitumen is stored in a mild steel tank of 30 m^3. This capacity can cope with a daily consumption of 3 m^3 with one weekly supply via a tank truck.

The tank is equipped with : a steam heating system, inlet and outlet connections and a level indicator.

The molten bitumen is kept in internal circulation by means of a gear-pump (capacity 5 m^3/h).

The hot bitumen is delivered to the extruder by means of a dosing gear-pump, type MAAG, with a max. capacity of 200 l/h.

A pressure reducer allowing max. 7 kg/cm^2 has been inserted in the supply line of the steam heating system to prevent excessive heating.

8.3 Extruder decontamination

Decontamination of the extruder is made by three successive operations :

- feeding of pure bitumen to the extruder during ca. 20 min.,
- washing with an organic non-burnable solvent to remove the bitumen residues,
- rinsing away the inorganic residues with water.

Washing with solvent was tried when the extruder outlet was plugged. Even after introduction of ca. 20 l solvent, the extruder outlet could not be unplugged.

8.4 Solvent recuperation

In the evaporator 46 the organic solvent is distilled off. The distillate is stored and can be recycled when necessary. After a few evaporation tests, made with evaporator 46, it appeared that :

- a poor steam supply control, resulted in a bad quality of the distillate,
- a density indication is worthwhile,
- the space under the evaporator was unsufficient for an easy emptying of the residual material,
- rerouting of the steam and condensate pipes in the vicinity of the evaporator was necessary to create enough space for shielding purposes.

Once the steam supply was equipped with a pressure control valve, some new tests were made.
The quality of the distillate was improved and the evaporation rate was easy to control.

Experimental data on the evaporator feed and distillate	
Evaporation rate l/h - water	0.07 - 0.9
- perchlorethylene	0.25 - 3.6
Concentration in aqueous phase :	
$NaNO_3$ g/l - feed	20.7 - 256
- distillate	0.02 - 1.04
Cl^- mg/l - feed	70 - 780
- distillate	3 - 9

Deductions from the table :

- the ratio water/perchlorethylene in the distillate is ca. 3.5,

- the reached decontamination factor varies between 80-250.

8.5 Extruder distillate routing and oil filtering

The aqueous distillate leaving the extruder condensors is collected in vessel 28. The oil fraction is separated from the water on the oil-filter 29.

The proposed filter material "Ekoperl" has an oil absorption capacity of ca. 290 g/l.

Taking into account 0.02 wt-% oil per liter of distillate and an average distillate flow of 120 l/h, it results in 24 g oil/hour. Considering a filter with a 5 liter Ekoperl content, it will be loaded with oil each 2.5 days. The intention is to take bigger oil filters in order to have about 1 week loading time. To avoid too much maintenance, it is considered to use an air-lift, feeding the distillate directly on the oil filter, prior to be collected in vessel 28.

The two horizontal Moineau pumps in room 007 will then only be used to empty vessel 28 to the hot waste tank 31 or to the warm waste tank 30.

However, further tests are still necessary before a final conclusion can be reached.

8.6 Drum filling station

General

At the extruder outlet a steam heated ball valve is mounted to close the extruder outlet during decontamination of the extruder. Below the ball valve outlet a steam heated funnel of approximately 30 l capacity is placed to collect the bitumen during the time that the drum is changed.

During drum changing the outlet of the funnel is closed by means of a piston valve.

Under the funnel we have a circular table 37 turning on ball bearings, driven stepwise by an electrical motor, the table supports 6 removable drip-trays and can receive 6 drums of 220 l.

Funnel ventilation

The gases released from the drum under filling and from the funnel are sucked by means of a special filling station ventilation.

In the exhaust duct of the funnel, two non-burnable filters are placed. These filters can be removed and renewed with the help of the telemanipulator. A butterfly type manual valve in the exhaust duct can be remotely operated to adapt the ventilation rate to the needs required.

DESCRIPTION OF TESTS PERFORMED ON AUXILIARY EQUIPMENT

1. AUXILIARY DRUM HANDLING EQUIPMENT IN ROOM 027

This room contains an overhead crane for the removal of the two concrete blocks closing the hatches in the floor above cells 005 and cell 013 and for introducing and retrieving the emergency telemanipulators.

Some experiments were made to handle the telemanipulators in the intervention room 027 and to introduce the telemanipulator in the cells 005 or 013. As a result of these tests some modifications were carried out. Additional guiding was made above the two supports on which the two telemanipulators are hanging in their rest position. This additional guiding is required to facilitate the removal of the telemanipulator from its support and its reintroduction.
The translation speed on the X-Y axes of the travelling crane has been reduced to a maximum speed of 4 respectively 1 m/min.
A detailed operation manual was made describing the successive operations necessary for introduction or retrieval of a telemanipulator into or from the cell.

The handling of this auxiliary drum handling equipment in intervention room 027 is rather complicated and time consuming, however it is not an operation to be made frequently since it is only an emergency equipment.

2. PUMPS

Centrifugal pumps, type KSB, are used in the circuits for cooling and warm water.

Screw pumps, Moineau type, are used to transfer the low level effluents and the radioactive slurries.
The pumping bodies of the slurry pumps are housed in the concrete floor above the feed vessel, while their motors are situated above the floor level.

Special care is taken in connection with :

- seal flushing,

- priming, and

- rinsing of the pump body after use.

A programmator guarantees that always the same sequence is followed for operation of these pumps.

During the first tests made with slurry, it was discovered that the transferred solution had a lower solids/liquid ratio than in the feed vessel 10. This deviation in the solids/liquid ratio was created by settling in the overdimensioned suction pipe of the pumps.
A bad surprise was the high RPM needed to prime the pumps, resulting in a peakflow at start-up.
It is believed that by decreasing the diameter of the suction pipe, both negative effects will be eliminated or at least significantly reduced.

To increase the safety of the transfers, alternatives were considered, especially to avoid regular maintenance of the slurry pumps. These alternatives (i.e. steam-jets and air-lifts) will be tested simultaneously with the pumps during the cold tests.

The first attempts to transfer slurry with the steam-jets turned out to be successful.

The double stage air-lift system chosen as alternative for feeding the extruder needs still some adaptation. Here, it was discovered an increase in the solids/liquid ratio between the slurry in the feed vessel 10 and the slurry transferred to the extruder, due to a settling in the first stage air-lift pot. Modification of the geometry of the pot is now considered, in view of reducing the risk of settling.

3. STIRRERS

The liquid make-up vessels, the reaction vessels 7 and 8, and the feed vessel 10 are equipped with a mechanical stirrer retrievable from the vessel through the top opening.

An abnormal vibration of the stirrers in the liquid make-up vessels is caused by their excentric installation. Supplementary gaskets were needed to improve the tightness at the flange connection of vessels 7, 8 and 10.
In vessel 10 it was discovered that the dead volume under the lowest propeller was 1300 l and the minimum mixing volume ca. 2 m^3.
To reduce this, the lowest propeller has been lowered to the end of the shaft, resulting in a dead volume decrease to 250 l.

4. INSTRUMENTATION

4.1 General

The main measurements required are : level, density, flow, pressure, gamma activity, conductivity and temperature.

In general, air-purged diptubes are used for level and density measurements inside active cells.

Electro-pneumatic transmitters, mostly Foxboro type, are located in a transmitter gallery which is situated above the liquid level in the vessels.

All diptubes between transmitters and vessels have an adequate slope of more than 5 %.

For temperature measurements, NiCr-Ni retractable thermocouples are used.

Nearly all recording, indicating and controlling instruments are centralized on the control panel in the control room 016.

4.2 Summary of the gained experience

a) Level and density measurement based on diptubes technique has proven to be very reliable. An accuracy up to 3-5 % can be guaranteed.
One of the main problems with diptubes is the risk of plugging due to crystallization.
To avoid this, all suspected ones have been provided with a humidifier pot so that humidified air is purged through. In spite of this precaution, frequent plugging of the diptubes occured in the vessel 10. By lowering the stirrer propeller in this vessel and thus decreasing the dead volume, it is hoped to eliminate this problem.

b) Level detection system in the drum filling station

The drum under filling, as well as the funnel are both equipped with two independent level detection systems. The level in the funnel will be detected by means of :

- ultra sonic detector,

- thermocouple.

The level in the drum will be detected by means of :

- gamma detector at two different levels,

- optical system (mirror + visual observation from the control room).

Some of the proposed systems are still under construction, the others are under testing.

c) Liquid, steam and gas-flow in pipelines are mostly measured by the differential pressure method.

Flow through a suitable primary device, inserted in lines, produces differential pressure inferential of flowrate.
The slurry feed flow, however, is measured with a magnetic flow meter.
Abnormal variations in the slurry feed flow measurement were observed, originating from :

- external influences on the magnetic flow meter (e.g. operation with H.F. welding equipment in the vicinity of the meter),

- liquid level fluctuations in the pipe passing through the magnetic flow meter.

As alternative the slurry feed flow can be deduced from :

- the calibration curve of the slurry pumps or air-lift,

- the depletion in tank 10, and

- the distillate flow.

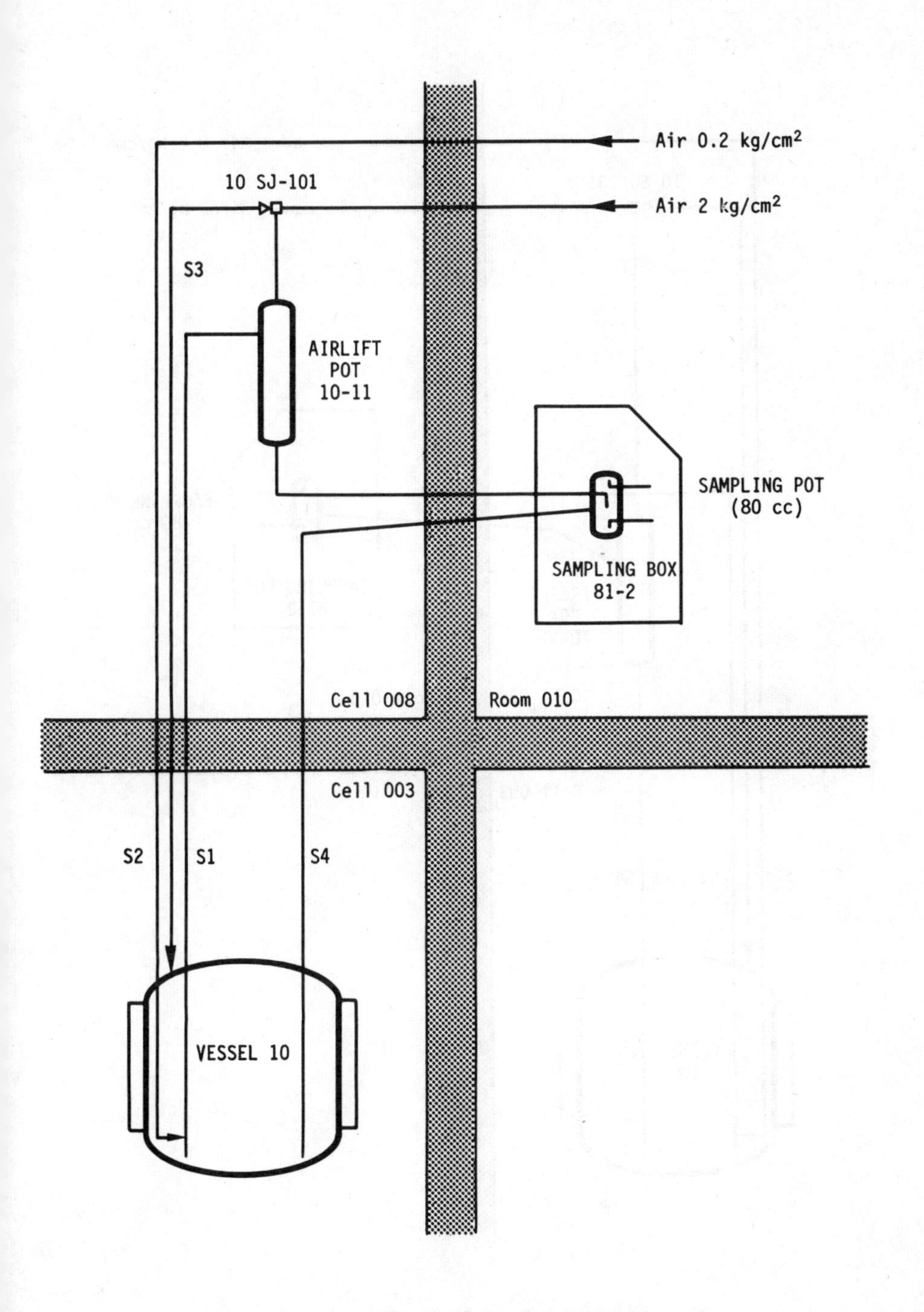

Fig. 1 - ORIGINAL SLURRY SAMPLING SYSTEM

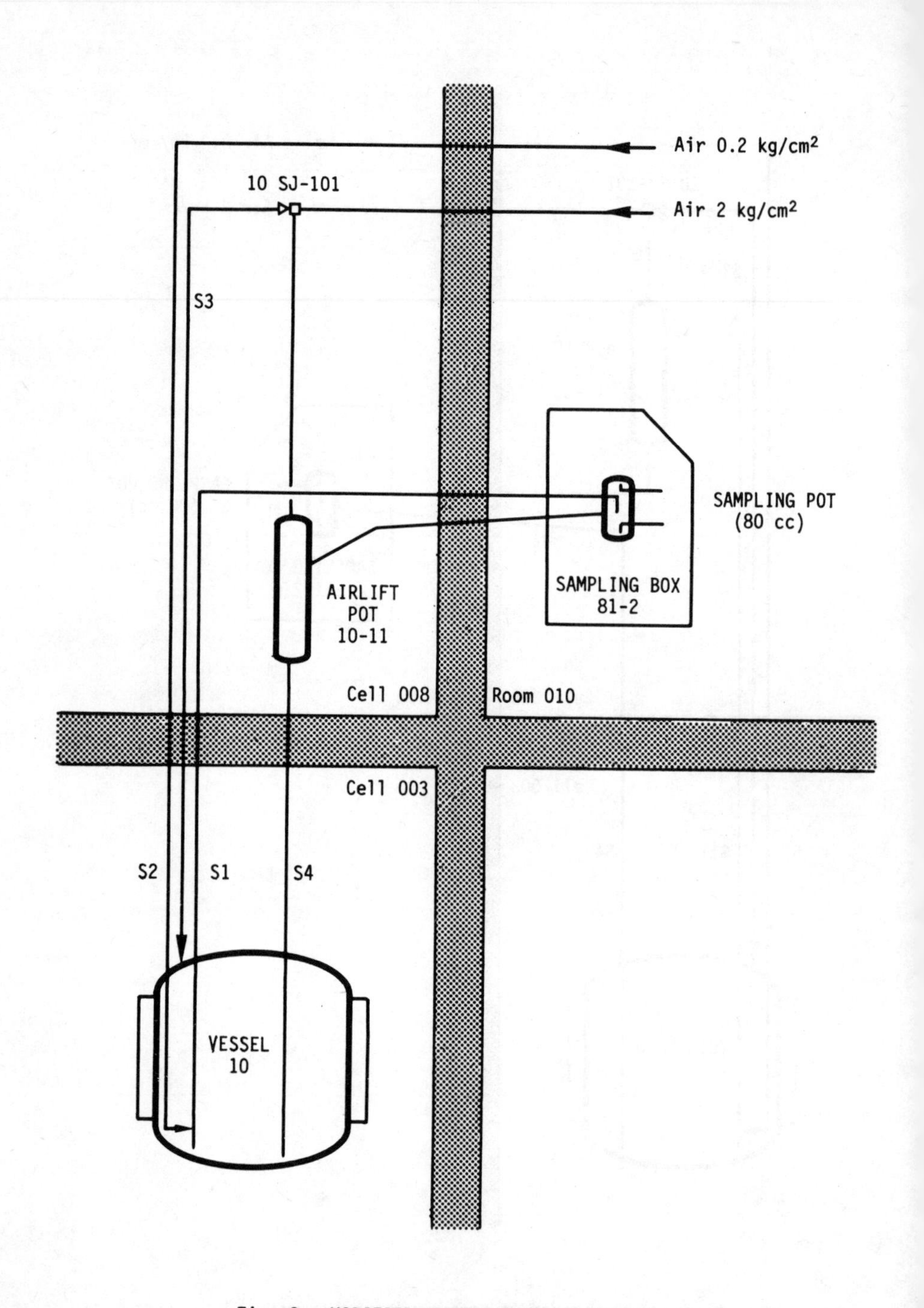

Fig. 2 - MODIFIED SLURRY SAMPLING SYSTEM

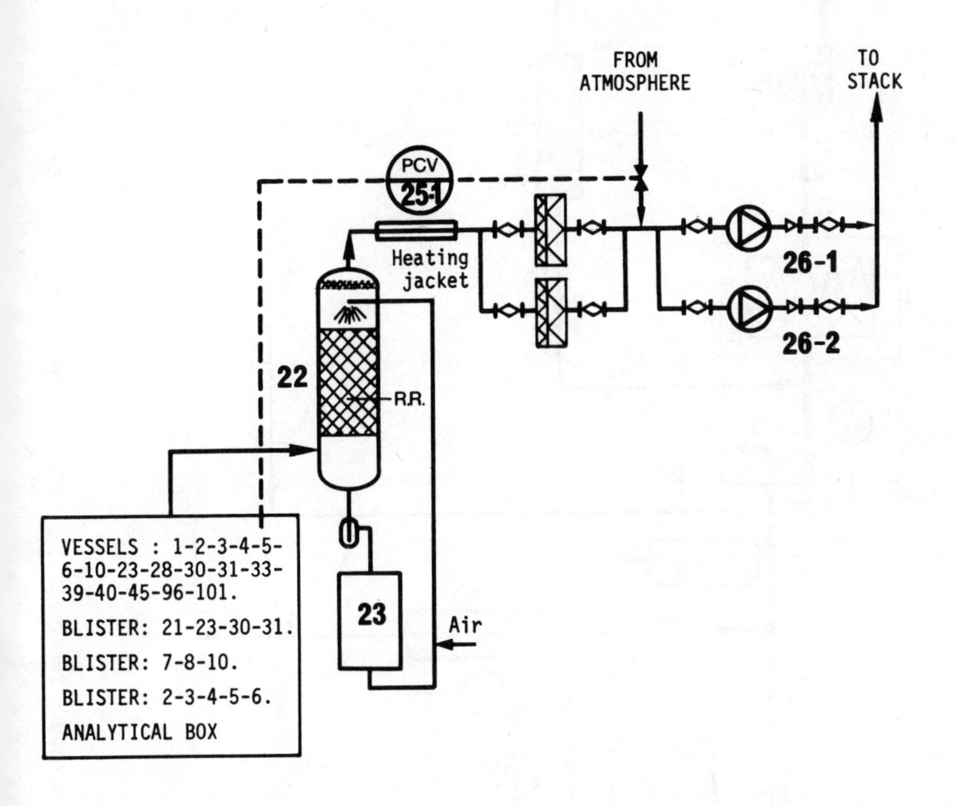

Fig. 3 - SCHEMATIC DRAWING OF THE NORMAL VESSEL VENTILATION SYSTEM AS TESTED

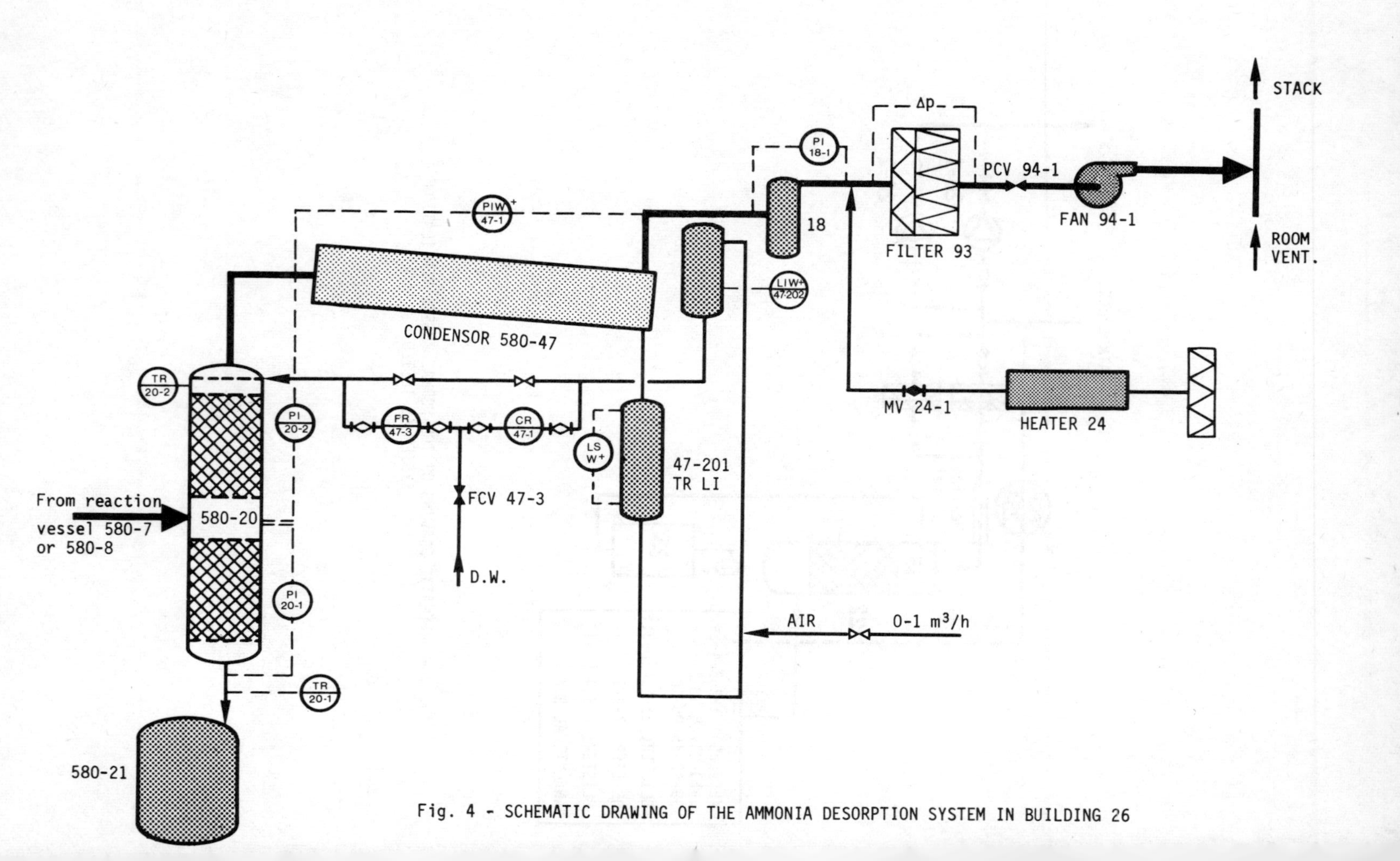

Fig. 4 - SCHEMATIC DRAWING OF THE AMMONIA DESORPTION SYSTEM IN BUILDING 26

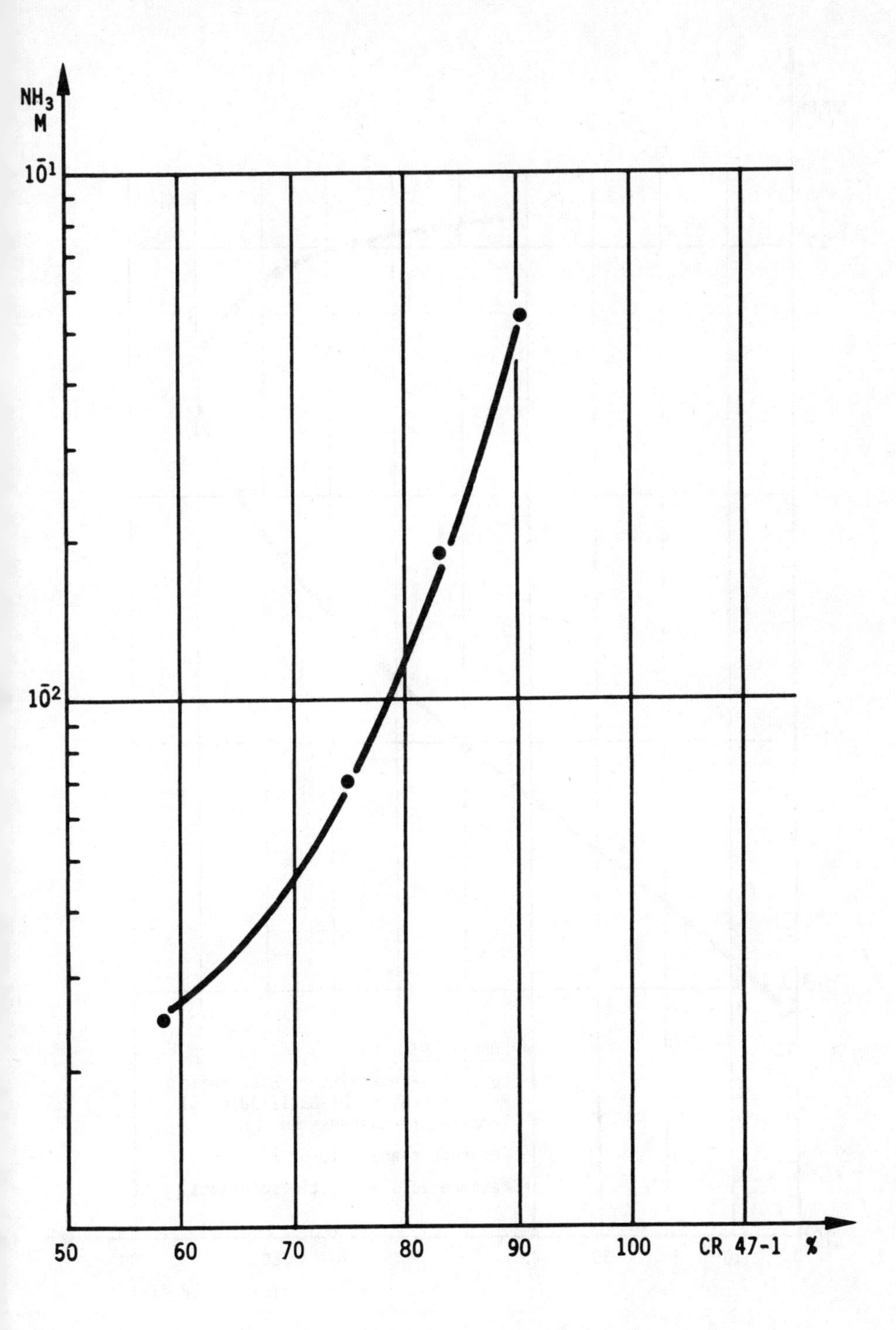

Fig. 5 - AMMONIA CONCENTRATION IN VESSEL 7 AS FUNCTION OF THE READING ON THE CR 47-1

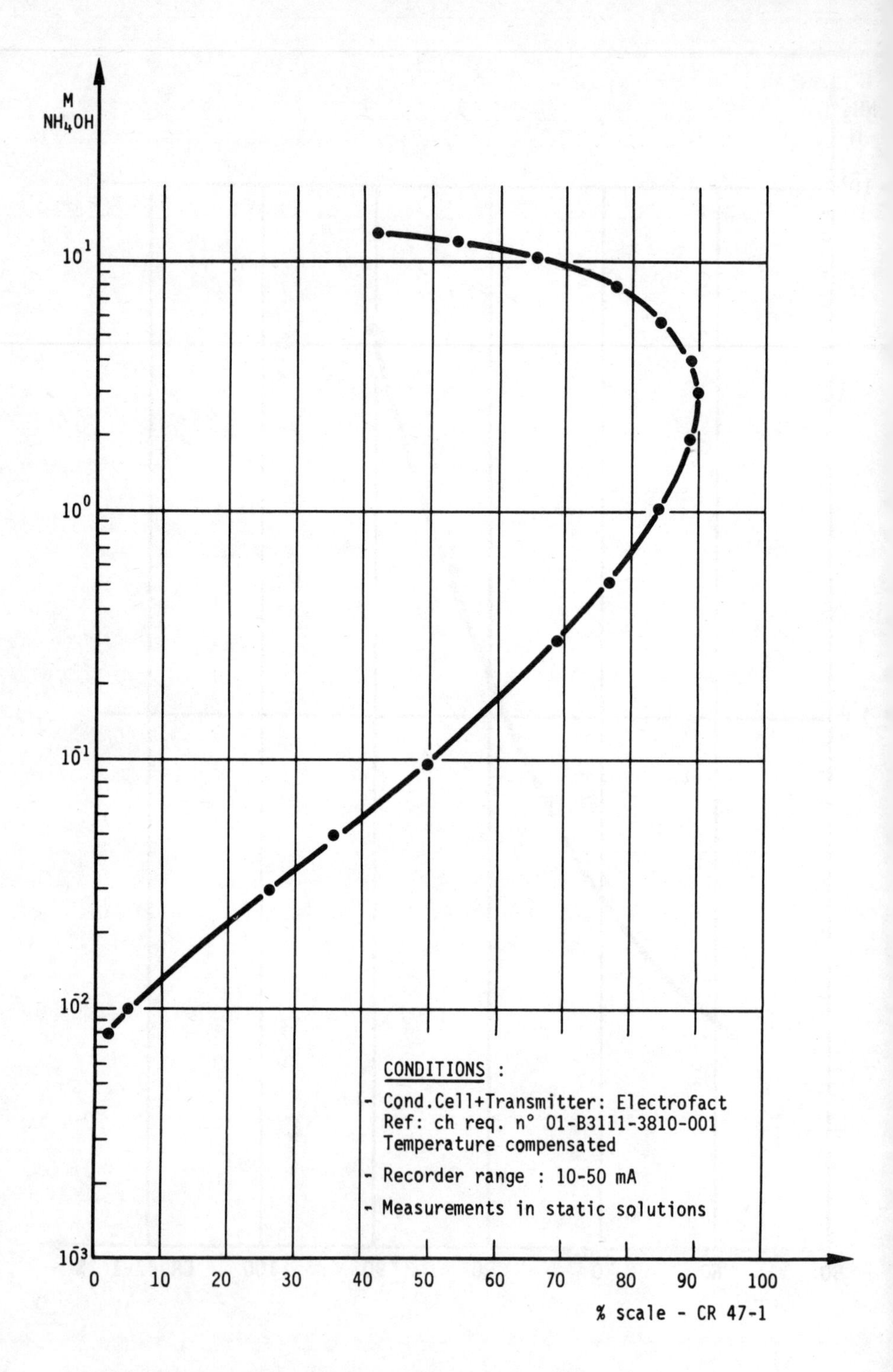

Fig. 6 - NH_4OH CALIBRATION ON CONDUCTIVITY RECORDER OF BLD 26
(calibration date : 17/12/75)

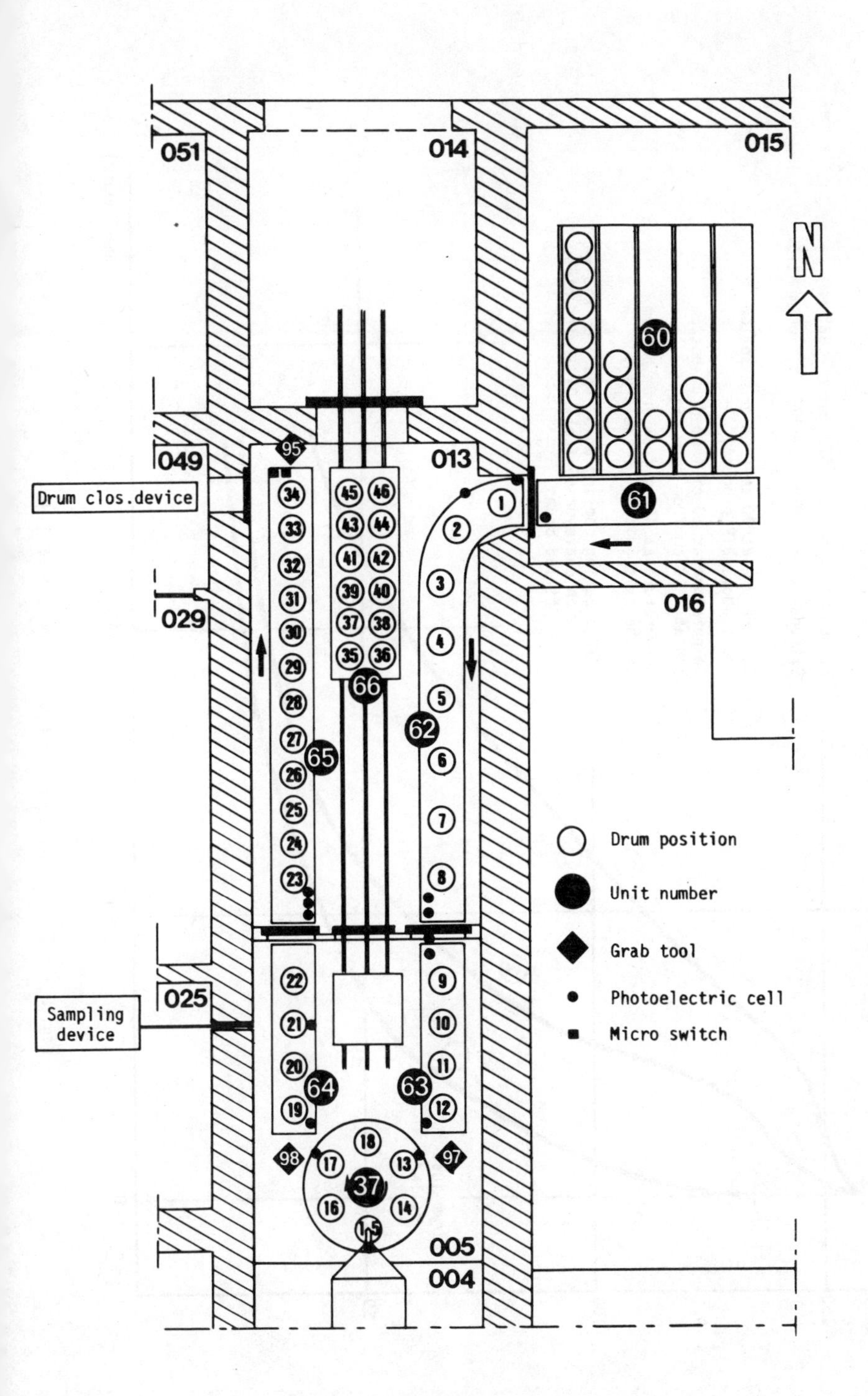

Fig. 7 - DRUM CONVEYING SYSTEM (BLD 26)

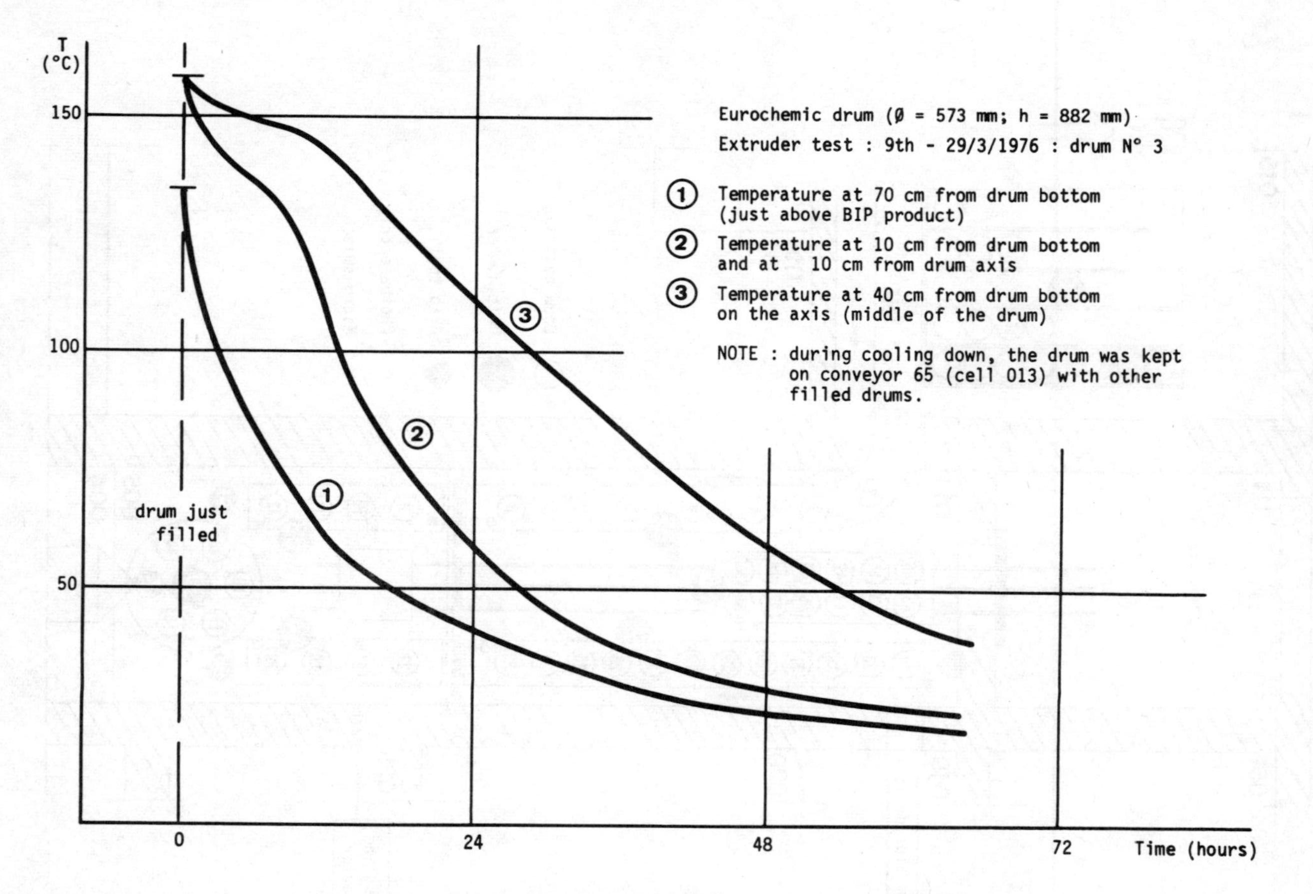

Fig. 8 - BITUMEN PRODUCT TEMPERATURE AS FUNCTION OF TIME

STORAGE FACILITY FOR SOLID MEDIUM LEVEL WASTE AT EUROCHEMIC

M. Balseyro-Castro

EUROCHEMIC

Mol, Belgium

ABSTRACT

An engineered surface storage facility is described ; it will serve for the interim storage of solid and solidified medium-level waste resulting from the reprocessing of irradiated fuels.

Up till now, two storage bunkers have been constructed. Each of them is 64 m long, 12 m wide and 8 m high and can take up to about 5,000 drums of 220 l volume.

The drums are stored in a vertical position and in four layers.

The waste product drums are transported by a wagon to the entrance of the bunkers from where they are transferred into the bunker by an overhead crane which is remotely controlled by high-frequency modulated laser beams.

A closed-circuit camera is used to watch the handling operations.

The waste stored is fully retrievable, either by means of an overhead crane or a lift-truck and can then be transported to an ultimate storage site.

1. INTRODUCTION

Liquid medium-level wastes (MLW), resulting from the reprocessing of irradiated fuels, are stored on the Eurochemic site. These waste solutions will be incorporated into blown bitumen and poured into 220 litre drums which are only filled to 80 %. The bitumen-waste product contains about 55 wt-% of bitumen and 45 wt-% of solids. The drums are made of "chromized" steel having very good corrosion resistance.

The bituminized waste will intermediately be stored at the Eurochemic site in a bunker building, called EUROSTORAGE.

In addition, solid waste of the same activity level and packed in a similar manner, coming from another source, can be handled and stored in this installation.

2. STORAGE FACILITY

2.1 The following principles have been taken into consideration :

- the storage facility is engineered and built to last for a period of at least 50 years, and is not considered as a final storage place.
- The waste stored must remain fully retrievable in view of its later transportation and disposal.

3. DESCRIPTION OF THE BUILDING

3.1 General lay-out (Figure 1)

In the prolongation of the axis of the cells 005, 013 and 014 of building 26, a liaison corridor 03 serves the storage cells to be built successively on the eastern side first, on the western side afterwards.

The transfer station 04 for the reception of the solid waste other than the drums coming from the bituminization installation in building 26, is situated at the beginning of the western side of the corridor.

Some annexes will house the auxiliaries, such as the ventilation fans and filters, electrical switchboards, etc.

3.2 Construction principles

The building is constructed in reinforced and partly in prestressed concrete. The wall, floor and roof thicknesses are determined by the more stringent requirements of either the building resistance or the shielding against radiation.

Special care was exercised to ensure the leak-tightness of the building during the whole of its expected lifetime.

A flexible sheet is imbedded in the floor rising in the walls in order to create a leak-tight "drip-tray".

The leak-tightness of the roof is ensured by multilayer coatings.

3.3 Control of the design

Due to the long-term integrity and leak-tightness required for these large buildings, all drawings and calculations of the architect-engineer have been verified and approved by an independent body of construction experts belonging to the University of Liège.

3.4 Finishing

The floors are made of a hard concrete layer in all cells and rooms except in the control room where the floor will be covered with PVC.

The walls are of painted concrete in all rooms.

The bunker cells are of bare concrete.

The ceiling is equipped with lighting fixtures in the control room.

3.5 Access

An access road will be built for the transfer station where the waste coming from other sources than building 26 will be received. The trucks will enter the transfer station 04 through a large door on the west side. Place for a shielded door is foreseen for later installation, if necessary.

Two entrances for the personnel are provided at floor level, one on the west side, the other on the east side of the corridor. These doors give direct access to staircases (06 and 08, one on each side) leading to the control room on the first floor level.

The door on the west side also gives access to the observation room 05 of the transfer station 04 while the door on the east side also gives access to the auxiliaries room 09.

3.6 Storage bunkers

Each storage bunker is designed to have a capacity of about 5,000 drums, which represents about 20 months'production.

The dimensions are approximatively :

length : 64 m
width : 12 m
height : 8 m

The floor is carefully levelled in order to obtain orderly stacking of the four layers of drums.

A network of small "ditches" collects any water entering the cell and brings it to an inspection pit.

An opening in the wall between the storage cell and the liaison corridor allows the passage of the overhead travelling crane. A chicane avoids direct radiation beam into the corridor from the stored drums.

When the storage cell is filled with drums, the opening is remotely closed with concrete blocks.

In the first stage, two storage bunkers have been built. A third storage bunker has to be built before any drum is loaded into the second one, and so on, in order to keep the exposure of the construction workers at the level acceptable for non-nuclear workers.

4. HANDLING EQUIPMENT

4.1 Handling requirements

The handling equipment is primarily intended for the transport of the bitumen waste product drums from the exit of building 26 to one of the storage bunkers, where the drums are piled up.

Due to the fact that these drums are not tightly closed (in order to allow the escape of possible radiolysis gases) they are transported and stored in a vertical position.

The dimensions of the product drums are as follows :

diameter	: 0.6 m
height	: 0.9 m
opening diameter	: 0.31 m
maximum weight	: 400 kg

The maximum product weight in a drum should not exceed 400 kg. Also as a design basis, it is considered that a maximum production of 24 drums per 24 hours (3 x 8 hours shift work) have to be handled and stored during the 7 useful hours of a normal workday.

The auxiliary handling equipment is designed for the solid waste coming from other sources than the bituminization facility. It can unload a transport truck after having removed the shielding covers and introduce the waste containers into the main handling chain.

4.2 Handling in the storage bunker

An overhead travelling crane (bunker crane) circulates in the storage bunker. The railing of this crane is extended to the west side of the liaison corridor which allows the bunker crane to travel in the corridor in front of the storage cell. This crane is equipped with a telescopic hoisting system, and it is electrically powered and fed from bus-bars installed in the storage cell. Two cranes with a capacity of 2 x 5 tonnes each are installed in the liaison corridor and they can travel between the transfer station and the end of the north side of the liaison corridor.

When a storage bunker is full, the two cranes are used to transfer the travelling crane to the next (empty) storage bunker.

The movements of the travelling crane are remotely controlled by high-frequency modulated laser beams. One set of lasers is located in the west side of the liaison corridor, in front of the storage bunker opening. The purpose of these lasers is to transmit the orders given to the overhead travelling crane and a feed-back for the respective movements of the handling equipment (x, y, z and grab tool).

These movements of the travelling crane can be automatically controlled from the picking-up of its load in the liaison corridor to the preselected positioning on the heap of drums. The lowering of the load and the movement of the grab tool are not automatically controlled. The automatic control also directs the movements of the crane passing the chicane established at the entrance of the storage bunker to prevent direct radiation beams coming from the stored drums.

Manual control of all these operations is also possible.

In the case of a drum laying in a horizontal position in the bunker (e.g. due to a maloperation), a special grab tool exists for handling the drum and bringing it to the right position.

Closed circuit television cameras and lighting devices installed on the overhead crane allow the operator to watch the operation on television screens installed in the control room.

Also, a television camera is installed on the emergency crane which can also travel in the storage bunker.

For the television cameras mounted on the overhead and emergency cranes, an additional laser set is used. All this laser information is transmitted to the control room via coaxial cables.

The drums are stored in a vertical position in four layers. Each layer is shifted a half-pitch (the pitch is 625 mm) in both directions, with regard to the layer immediately below :

1st layer	15 x 94 =	1,410 drums
2nd layer	14 x 93 =	1,302 drums
3rd layer	13 x 92 =	1,196 drums
4th layer	12 x 91 =	1,092 drums
		5,000 drums.

An important space (about 4 m) is kept free between the upper layer of drums and the roof for the movement of drums suspended from the grab-tool of the travelling crane. The lifting capacity of the travelling crane is 2 tonnes which gives an important margin above the maximum weight of a drum filled with a bitumen waste product mixture. This margin allows for the handling of drums filled with solid waste imbedded in concrete or four drums at a time filled with bitumen waste product mixtures.

4.3 Handling in the liaison corridor

A transfer carriage loaded with 12 drums in cell 013 of building 26 travels in the corridor until in position in front of the storage cell, where the drums are picked up by the travelling crane.

The daily capacity of 24 drums requires two loading-unloading cycles for the transfer carriage.

The transfer carriage is electrically powered and fed from bus-bars installed in the corridor. The movements of the carriage in cells 013 and 014 of building 26 are controlled from the control room of building 26 (bituminization facility). The movements of the carriage in the liaison corridor are controlled from the control room of building 27.

4.4 Handling in the transfer station

The solid waste coming from other sources than building 26 is shipped by trucks which are introduced at ground floor level in the transfer station 04.

An overhead travelling crane is installed in room 04 at first floor level and circulates on rails extended in the liaison corridor. Through a hatch in the floor, it can remove the truck shielding with a first hoist having a capacity of 5 tonnes, and pick up the waste drum with a second hoist having a capacity of 2 tonnes.

The waste drum is then transferred to the liaison corridor and is deposited on a transfer carriage, similar to the carriage used for the bitumen drums, running on a parallel line in the liaison corridor. The transfer carriage brings the waste drums in front of the storage cell, where it can be picked up by the storage cell crane.

The transfer station overhead crane is controlled from an auxiliary control room 05 in the transfer station or from the main control room, with a direct view into the liaison corridor through a lead-glass window.

A television camera is also mounted on the overhead crane of the transfer station to allow the operator to watch the crane movements from the operation desk of the control room.

4.5 Safety devices

Various electrical interlocks are built in the control systems of the handling equipment in order to prevent accidental maloperations leading to equipment damage.

Examples :

a) the overhead travelling crane cannot be moved if the load is not in the upper position ;

b) the telescopic column of the hoisting system of the travelling crane is equipped with safety switches to stop the crane moving in case of maloperation during travelling in the chicane area ;

c) the grab tool of the storage bunker crane cannot be closed if the drum is not at the right position ;

d) the transfer carriage cannot be moved if the adequate door in building 26 is not open ;

e) a switch allows only one control room at a time to operate the transfer carriage ; etc.

4.6 Maintenance

The maintenance of the handling equipment is done in the liaison corridor, which must be free of any waste drums before it can be entered by the maintenance crew.

For maintenance, the overhead crane of the storage bunker is transferred to the front of the empty bunker or to the end of the liaison corridor to avoid direct radiation exposure to the personnel.

4.7 Emergency situations

Means have been developed in order to cope with every imaginable emergency situation without radiation exposure to the personnel.

Examples :

a) a transfer carriage, loaded with drums, and stalled midway in the liaison corridor, can be remotely pulled, either to the position in front of the storage cell, where the crane can remove the drums, or back to building 26 ;

b) if the overhead travelling crane has stalled in the storage bunker, it can be brought back remotely to the liaison corridor for repair ;

c) the movements of the overhead crane of the transfer station can be done manually and remotely, except the hoisting which is equipped with a second motor and fed by a separate circuit.

5. AUXILIARY EQUIPMENT

5.1 Ventilation - Heating

5.1.1 Storage cells and liaison corridor

The conditioning of the radioactive material handled and/or stored in these rooms is such that the material is very unlikely to produce any loose contamination and is still more unlikely to produce any airborne contamination.

Therefore, slight ventilation is only required for the evacuation of hydrogen produced by the radiolysis of bitumen.

Calculations, based on conservative assumptions, show that it requires only 16 m^3 of fresh air per day to keep the hydrogen concentration below the lower explosive limit, for the maximum activity content in one storage bunker. It represents about one renewal per year in the free space of the storage bunker. Such a low renewal rate can be ensured by natural circulation of the air if some openings are present.

However, as practical experience with such storage conditions is still rather limited, it has been installed with a low-flow forced ventilation system, with filters at the inlet and at the outlet.

Fresh air, drawn from the outside through filters, will be supplied to the liaison corridor. The air is extracted in the lower part of the liaison corridor and in the upper part of the storage cells, hydrogen being a light gas. The extraction is provided for by fans and absolute filters.

The balance of supply and exhaust will be regulated in such a way as to ensure slight underpressure in the storage cells.

Cell 014 will serve as an airlock separating the ventilation systems of buildings 26 and 27.

5.1.2 Normally occupied rooms

Rooms which are normally permanently or intermittently occupied need a ventilation-heating system. This is fully independant of the system described in 5.1.1 and supplies fresh air, if necessary heated and conditioned, to the control rooms and to the rooms housing the auxiliary equipment.

The main control room will be kept at a slight overpressure towards the outside.

5.2 Utilities

5.2.1 Electricity and lighting

The switchboard for building 27 will be fed with low tension from the switchboard of building 26. The total power required for building 27 is about 45 kVA.

6. SAFETY ASPECTS

6.1 Nuclear Safety

6.1.1 Criticality

The concentrations of fissile materials (uranium and plutonium) in the solid wastes to be stored are so low that any criticality risk can be completely excluded.

6.1.2 Contamination

The nature of the materials and the quality of the containment make surface or airborne contamination very unlikely during handling and the first years of storage.

Fixed monitoring equipment is not foreseen.

Important ingresses of water can only result from a leaking roof. Water entering in the storage cells will be collected by a network of drains in the floor and brought to collection pits where it will be detected. Samples can be taken and measured for activity. Remedial actions will have to be taken in order to restore the leak-tightness of the building. Airborne contamination will not be created by a slow flow of air circulating between undisturbed material.

The material of the drums and the conditions of storage are selected with the aim of ensuring good conditions for the retrieval operation.

If the resistance of the drums is reduced by unexpectedly severe corrosion, the handling of the bitumen blocks with a view to repackaging is feasible. On the most pessimistic and very unlikely assumption of completely corroded drums and divided bitumen blocks, the design and leak-tightness of the buildings will allow the possibility of organizing retrieval and reconditioning operations in a safe manner.

6.1.3 Irradiation

The thickness of the walls of the storage cells has been calculated in order to obtain the following exposure rates in contact with the outer surface of the wall :

roof	2.5 mrem/h
exterior wall	0.25 mrem/h
wall towards the next cell	2.5 mrem/h
wall towards the corridor	0.25 mrem/h

These values are obtained in a cell filled with 5,000 drums containing a bitumen-waste product mixture having a specific gamma activity of 0.5 Ci/l at 0.7 MeV per disintegration.

For the transfer station 04, an exposure rate of 2.5 mrem/h in contact with the outer surface of the wall is considered acceptable due to the limited time during which transfers are performed. This value is calculated for one drum of 200 l containing waste having a specific gamma activity of 1 Ci/l at 0.7 MeV.

Fixed gamma radiation monitors, with recording and alarm, will be installed in the control room and in the observation room.

Auxiliary gamma monitors will be installed in the transfer station and in the liaison corridor, in order to have an indication of the presence of a radiation source.

Portable radiation monitors will be used for surveillance, according to the operation needs : interventions, maintenance, etc.

6.2 Conventional Safety

The conventional safety of normal industrial equipment such as electrical supply and distribution, compressed air distribution, steam heating, lifting devices, etc. is ensured in full compliance with the regulations of the Belgian "Règlement Général pour la Protection du Travail".

6.2.1 Protection against fire

Apart from the insulation material in the electrical equipment, the only burnable material present in important amounts is the bitumen waste product mixture contained in metallic drums.

It cannot be considered as a major fire hazard : bitumen must be heated above 300°C before it will burn, if a source of fire is present.

The total decay heat produced in a storage bunker (5,000 drums) amounts to a maximum of 2.7 kW which is negligible compared to the mass of the building and the dissipation surfaces, and will not heat the bitumen to such temperatures.

It is not foreseen to install fire detectors or fixed fire fighting equipment in any part of the building. Only portable fire extinguishers will be provided in the control room and the auxiliaries room.

A conceivable fire risk for the stored bitumen waste product mixture would be the accidental crash of a military airplane, producing a large hole in the roof through which burning kerosene could enter the cell.

However, the thickness of the roof (0.75 m) and the prestressed construction give sufficient resistance ; kerosene infiltrating through cracks would not burn when arriving inside the cell.

6.2.2 Explosion

Submitted to radiation, organic material evolves hydrogen by radiolysis.

It has been calculated that one drum will produce at maximum 25 cm^3 of hydrogen per day. For one storage bunker it is sufficient to renew 16 m^3 of air per day in order to remain below one fifth of the lower explosibility limit for hydrogen in air (4 vol-%).

The free volume in one cell is more than 4,000 m^3. The very small renewal rate necessary is ensured by the natural respiration of the building due to atmospheric pressure variations. In addition, hydrogen being lighter than air will escape through the exhaust openings provided in the upper part of the cell.

Nevertheless, forced ventilation with filters will be installed, at least for the first storage cells, until some experience is gained on the long term behaviour of bituminized waste.

The occurrence of an explosive hydrogen-air mixture is thus very unlikely.

During the loading of the cell, electrical sparks can come from the operation of the handling equipment but at that moment, wide openings and movements of loads make the occurrence of an explosive mixture practically impossible. During the storage period, no source for electrical sparks is present in the cell.

However improbable, an explosion would probably not create major damage in the very large volume, thick-walled cell.

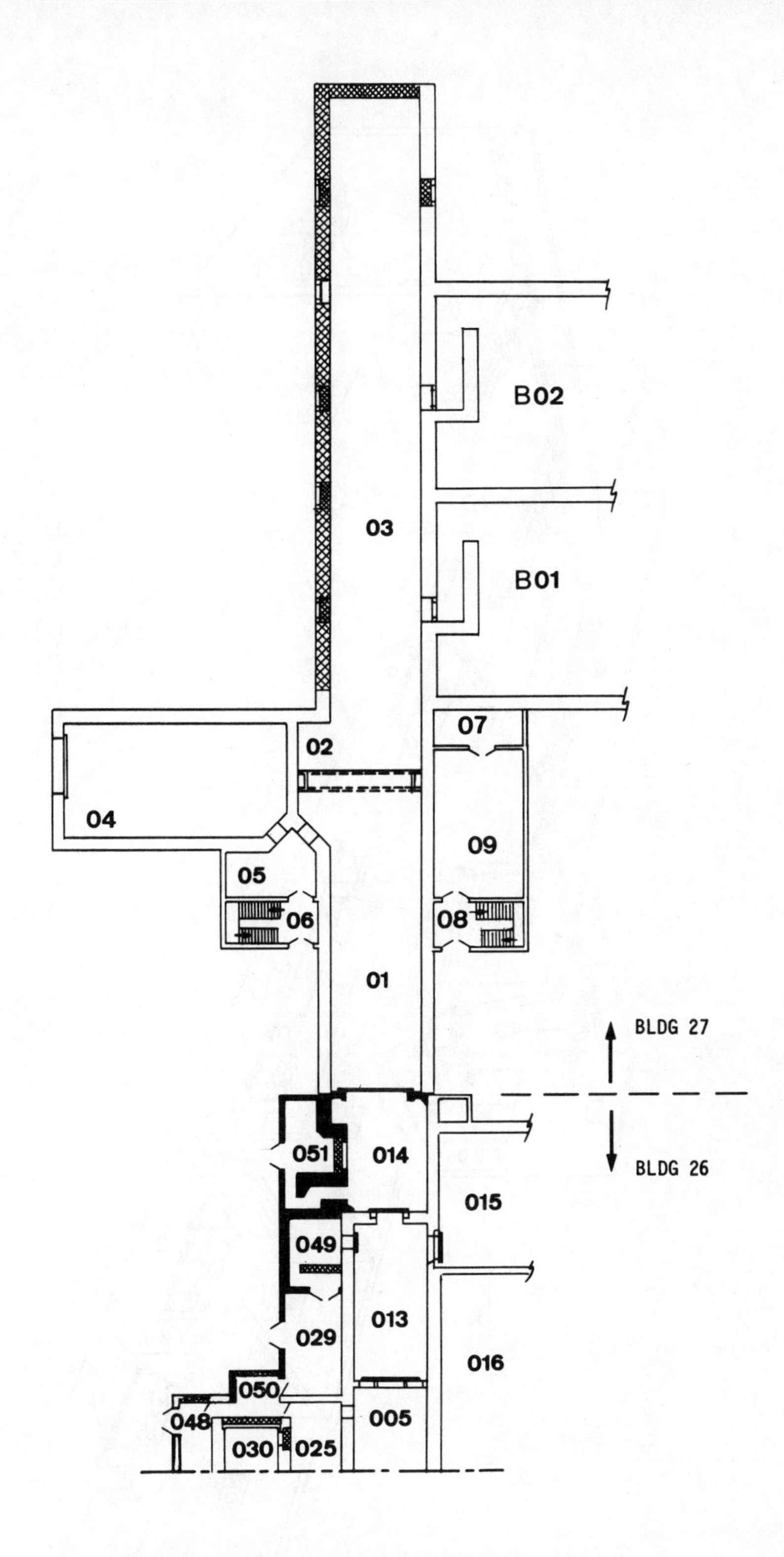

Fig. 1 - PLAN-VIEW OF BUILDING 27 (Ground floor)

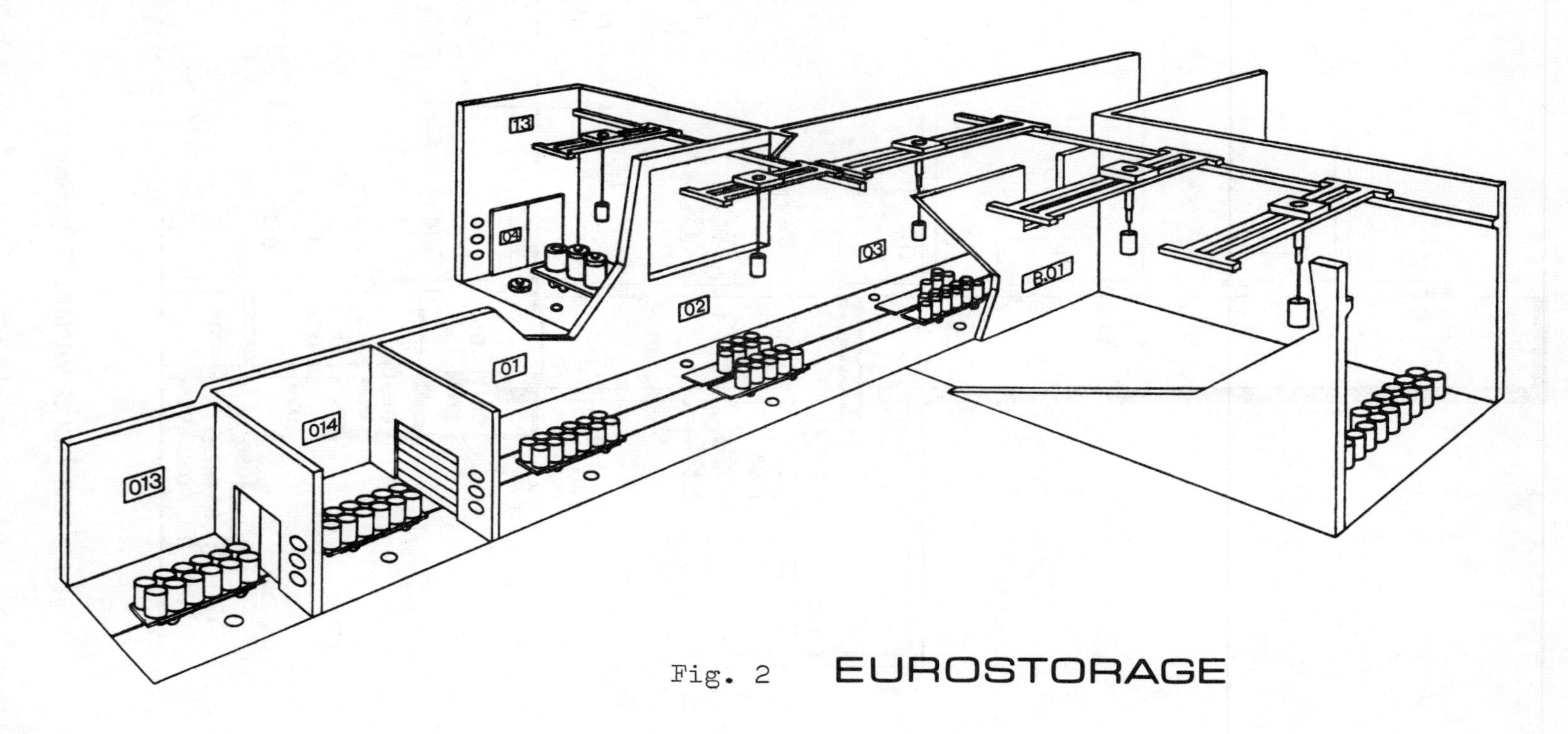

Fig. 2 EUROSTORAGE

Discussion

H.R. BURRI, Switzerland

How many air changes will there be per hour in the storage building ?

M. BALSEYRO CASTRO, Eurochemic

The air renewal will be in the order of 1.2-1.4 times per day, depending on whether the storage bunker is empty or completely full.

H.R. BURRI, Switzerland

Why are the drums stored only four layers high ?

M. BALSEYRO-CASTRO, Eurochemic

The telescope-type bunker crane moves inside the bunker at a height corresponding approximately to the height of a fifth layer of drums. Therefore the filled product drums are stored only in four layers but a fifth layer could still be stacked. The first reason is to avoid a drum being handled by the bunker crane striking the highest layer of drums due to an error in operation. The bunker crane does not start its movement if the telescopic column is not in the highest travelling position. The second reason is to be able to pick up a drum at any position in the bunker, which would not be possible if the drums were stored in more than four layers.

H.R. BURRI, Switzerland

What means have been provided to retract a crane e.g. in case of mechanical failure ?

M. BALSEYRO-CASTRO, Eurochemic

The bunker crane can be pulled out from the storage bunker, in case of mechanical or electrical failure, by means of a simple bridge which can be introduced into the bunker, having a separate power feeding and which is able to pull the stalled crane out of the bunker.

E. DETILLEUX, Eurochemic

1. Ventilation of the storage building

The problem of ventilation was extensively discussed during the study of the project. The need for ventilation is justified mainly by the risk of release of radiolysis gases from the bitumen product.

The air flow which would be required for a full bunker (5000 drums) is 16 m^3 per hour maximum (dilution of hydrogen).

It is likely, if not certain, that natural ventilation would have been sufficient. The decision to install forced ventilation at 300 m^3 per hour in the first two bunkers was taken in order to study the problem.

2. Vertical stacking of the drums has been decided to take into account the risks of fluidity of the bitumen product, the drum not being leaktight to allow release of possible radiolysis gas. There was no expert prepared to run the risk of advising storage of the drums in a horizontal position.

H. DWORSCHAK, Italy

Have the buildings been designed and constructed to be safe during seismic events ?

M. BALSEYRO-CASTRO, Eurochemic

An earthquake situation was not considered in the project.

K. BRODERSEN, Denmark

Can you tell us, if you have any possibility of retrieving a misplaced drum, for instance, one which has fallen on the floor ?

M. BALSEYRO-CASTRO, Eurochemic

If a drum falls on the floor and remains in a horizontal position, a special grab tool can be attached to the telescopic column which has a rotating movement, which is able to grasp the drum, to tilt it and bring it to a vertical position.

P.W. KNUTTI, Switzerland

What is the maximum temperature of storage when the storage facility is full ?

M. BALSEYRO-CASTRO, Eurochemic

The total decay heat in a storage bunker (5000 drums) amounts to a maximum of 2.7 kW, which is negligible compared to the mass of the building.

P.W. KNUTTI, Switzerland

Is retrieval, of the first placed barrel, possible with a view to its final storage ?

M. BALSEYRO-CASTRO, Eurochemic

All the drums are retrievable by means of the bunker crane, if still in good condition, or with a shielded lift truck through a foreseen opening on the east wall of the storage bunkers, which is normally closed with concrete blocks.

P.W. KNUTTI, Switzerland

Why is the next layer up shifted by ½ pitch in both directions, does this not prevent good air flow (cooling effect) ?

M. BALSEYRO-CASTRO, Eurochemic

The purpose of this stacking procedure is for stability reasons only.

G. LEFILLATRE, Fránce

Given the storage time in the bunkers (50 years) have you devoted some attention to the following possibility : corrosion of a certain number of drums, sufficient to allow the bitumen product to spread on the ground due to its fluidity, which would result in a possible blockage of the stacks of drums. What would you do in that case when you want to retrieve the drums ? Have you foreseen any suitable measures in the bunker and appropriate intervention equipment ?

M. BALSEYRO-CASTRO, Eurochemic

The characteristics of the drum material (chromized steel) are such that the bitumen mixture cannot come out of the drum as long as the outside and inside chromized layers remain, even if all mild steel is corroded, and hold the content of the drum. Nevertheless, the original shape could be modified, particularly for the lower layers.

E. DETILLEUX, Eurochemic

The weather resistance tests (storage outside) carried out for three years, show that the material chosen is satisfactorily resistant. The ambient conditions in the storage building are notably more favourable than those of the tests. We can therefore reasonably expect that the drums chosen will resist satisfactorily during the period foreseen for surface storage (about 50 years).

J.E. STEWART, United States

Have you made provisions for monitoring the surface radiation from each drum and any provisions for decontamination ?

M. BALSEYRO-CASTRO, Eurochemic

As far as monitoring surface radiation and decontamination are concerned, nothing was foreseen in this respect. Nevertheless, it is possible that there will be some spilling during the filling operation, mainly on top of the drum. This kind of contamination is practically fixed on the drum itself and does not produce any airborne contamination.

The radiation level at the surface of the drum can be determined from the sample taken from the bitumen mixture.

N. FERNANDEZ, France

Was the lifetime of the installation (50 years) foreseen to allow easy retrieval (without radiation protection problems) for transport to long term storage? Do you have any indication of the policy concerning long term storage ?

E. DETILLEUX, Eurochemic

Eurochemic, which is an international organisation situated in Belgium, has to comply with the requirements of the Belgian authorities concerning storage. The present measures make it possible to regard surface storage as an interim solution for a period of about 50 years. The long term policy is under study by the Belgian authorities.

Discussion

H.R. BURRI, Suisse

Combien de fois par heure l'air sera-t-il renouvelé dans le bâtiment de stockage ?

M. BALSEYRO CASTRO, Eurochemic

L'air sera renouvelé de l'ordre de 1,2 à 1,4 fois par jour, suivant que l'enceinte de stockage sera vide ou tout à fait pleine.

H.R. BURRI, Suisse

Pourquoi les fûts ne sont-ils stockés que sur quatre couches ?

M. BALSEYRO CASTRO, Eurochemic

La grue de type télescopique se déplace à l'intérieur de l'enceinte de stockage à une hauteur qui correspond approximativement à celle d'une cinquième couche de fûts. En conséquence, les fûts remplis de produits ne sont stockés que sur quatre couches mais on pourrait encore empiler une cinquième couche. Deux raisons incitent toutefois à ne pas le faire : En premier lieu, il s'agit d'éviter qu'un fût déplacé par la grue de l'enceinte de stockage ne vienne heurter les fûts de la couche supérieure par suite d'une erreur de manoeuvre. Ladite grue ne se met en mouvement que si la colonne télescopique n'est pas en bout de course. En second lieu, on veut pouvoir retirer un fût se trouvant dans n'importe quelle position à l'intérieur de l'enceinte de stockage, ce qui ne serait pas possible si les fûts étaient stockés sur plus de quatre couches.

H.R. BURRI, Suisse

Quels moyens ont été prévus pour retirer une grue, en cas de défaillance mécanique par exemple ?

M. BALSEYRO CASTRO, Eurochemic

La grue peut être retirée de l'enceinte de stockage, en cas de défaillance mécanique ou électrique, au moyen d'un simple pont alimenté en énergie par un système distinct, qui peut être introduit dans l'enceinte en question et est capable de retirer la grue bloquée de l'enceinte.

E. DETILLEUX, Eurochemic

1. Ventilation du stockage

Au cours de l'étude du projet, le problème de la ventilation a été longuement discuté. Le besoin d'une ventilation est essentiellement justifié par le risque de dégagement des gaz de radiolyse du bitume.

Le débit nécessaire, pour un bunker plein (5.000 fûts), est au maximum de 16 m^3/h (dilution de l'hydrogène).

Il est probable, et même certain, qu'une ventilation naturelle serait suffisante. C'est dans un but d'étude du problème qu'il a été décidé d'installer une ventilation forcée (300 m^3/h) dans les deux premiers "bunkers".

2. L'empilement vertical des fûts a été décidé en raison des risques de fluage de l'incorporé, la fermeture des fûts n'étant pas hermétique pour permettre l'échappement des gaz de radiolyse éventuels. Aucun expert, sur le bitume, consulté n'a pris le risque de conseiller un stockage avec empilement horizontal des fûts.

H. DWORSCHAK, Italie

Les bâtiments ont-ils été conçus et construits pour présenter toute sécurité en cas de secousse sismique ?

M. BALSEYRO CASTRO, Eurochemic

Le cas d'un tremblement de terre n'a pas été envisagé dans le projet.

K. BRODERSEN, Danemark

Pouvez-vous nous dire si vous avez quelque possibilité de récupérer un fût mal placé, et notamment un fût qui serait tombé par terre ?

M. BALSEYRO CASTRO, Eurochemic

Si un fût tombe par terre, et reste en position horizontale on peut fixer sur la colonne télescopique un grappin spécial à mouvement rotatif qui est apte à saisir le fût, à le retourner et à le placer en position verticale.

P.W. KNUTTI, Suisse

Quelle est la température de stockage maximale lorsque l'installation de stockage est pleine ?

M. BALSEYRO CASTRO, Eurochemic

La chaleur de décroissance totale dans une enceinte de stockage (5.000 fûts) représente 2,7 kW au maximum, ce qui est négligeable par rapport à la masse du bâtiment.

P.W. KNUTTI, Suisse

Est-il possible de récupérer le fût placé en premier lieu en vue de son stockage définitif ?

M. BALSEYRO CASTRO, Eurochemic

Tous les fûts peuvent être récupérés au moyen de la grue de l'enceinte de stockage, à condition qu'elle demeure en bon état, ou à l'aide d'un chariot de levage blindé à travers une ouverture prévue dans la paroi Est de l'enceinte de stockage qui est normalement fermée par des blocs de béton.

P.W. KNUTTI, Suisse

Pourquoi les couches sont-elles décalées les unes par rapport aux autres de la largeur d'un demi-fût dans les deux directions ? Cela n'empêche-t-il pas une bonne circulation de l'air (effet de refroidissement) ?

M. BALSEYRO CASTRO, Eurochemic

Ce procédé d'empilement n'est employé que pour des raisons de stabilité.

G. LEFILLATRE, France

Compte tenu du temps de stockage dans les bunkers (50 ans), avez-vous pensé à l'hypothèse suivante : Corrosion d'un certain nombre de fûts suffisante pour avoir un fluage de l'enrobé sur le sol, d'où il résulte un blocage possible de piles de fûts ? Que feriez-vous dans ce cas au moment de la reprise des fûts ? Avez-vous prévu des aménagements appropriés dans le bunker et un appareillage d'intervention adapté ?

M. BALSEYRO CASTRO, Eurochemic

Les caractéristiques du matériau constituant les fûts (acier chromé) sont telles que le mélange de bitume ne peut s'échapper du fût aussi longtemps que les couches chromées intérieures et extérieures restent en place, même si tout l'acier doux est corrodé, et qu'elles retiennent le contenu du fût. Néanmoins, la forme originale des fûts pourrait être modifiée, notamment au niveau des couches inférieures.

E. DETILLEUX, Eurochemic

Les essais de résistance aux intempéries (stockage à l'extérieur) conduits depuis trois ans montrent que le matériau choisi résiste de façon satisfaisante. Les conditions ambiantes dans le stockage sont nettement plus favorables que celles des essais. Il est donc raisonnablement permis de penser que les fûts choisis résisteront parfaitement pendant la durée prévue pour le stockage en surface (~ 50 ans).

J.E. STEWART, Etats-Unis

Avez-vous pris des dispositions pour surveiller le niveau des rayonnements à la surface de chaque fût, ainsi que pour assurer la décontamination ?

M. BALSEYRO CASTRO, Eurochemic

Rien n'a été prévu en ce qui concerne la surveillance du niveau de rayonnement à la surface et la décontamination. Néanmoins, quelques écoulements peuvent se produire pendant l'opération de remplissage, notamment en haut du fût. Ce type de contamination est pratiquement fixé à la surface du fût lui-même et n'engendre pas de contamination dans l'air.

Le niveau de rayonnement à la surface du fût peut être déterminé d'après l'échantillon prélevé dans le mélange de bitume.

N. FERNANDEZ, France

La durée de vie de l'installation (50 ans), a-t-elle été prévue pour permettre alors une évacuation facile (sans problème de protection) vers un stockage à très long terme ? Y a-t-il des éléments de politique au sujet du stockage à long terme ?

E. DETILLEUX, Eurochemic

Eurochemic, organisation internationale établie en Belgique, ne peut que se conformer aux directives des autorités belges en matière de stockage. Les dispositions actuelles permettent de considérer le stockage en surface comme une solution intérimaire pour une durée de l'ordre de 50 ans. La politique de stockage à long terme fait l'objet d'études par les autorités et les organismes belges.

Session II

Chairman - Président

P. DEJONGHE

Séance II

EXPERIENCE DANS LE DOMAINE DE L'ENROBAGE DANS LE BITUME EN BELGIQUE

N. Van de Voorde et K. Peeters.
S.C.K./C.E.N.
Mol (Belgique).

1. INTRODUCTION

L'enrobage de concentrats radioactifs dans une matière inerte en vue de préserver les substances actives de la lixiviation naturelle a toujours constitué l'objectif ultime de la plupart des méthodes de traitement des résidus. Au début des années 60, la méthode conçue à Mol était essentiellement basée sur le séchage de concentrats plus ou moins humides par évaporation dans un bain de bitume, ce qui avait pour effet de produire par la même occasion un mélange homogène bitume/concentrat. De plus, ce mélange pouvait être évacué à une certaine température du mélangeur-évaporateur. Ensuite, après refroidissement jusqu'à la température normale, le durcissement du mélange donnait lieu à une masse compacte, tout en gardant une certaine élasticité. D'une grande simplicité, cette méthode classique appliquée jusqu'à ce jour consistait donc à mélanger sous agitation une matière solide à un liquide plus ou moins visqueux, dans le but d'évaporer la plus grande partie de l'eau présente dans les concentrats, et éviter ainsi les difficultés inhérentes à l'utilisation de la concentration traditionnelle, c'est à dire la cristallisation des sels, la corrosion etc. L'installation qui est en service depuis plus de douze ans au département de traitement des résidus du C.E.N./S.C.K. à Mol a été spécialement conçue en vue du traitement des concentrats provenant du traitement des effluents faiblement actifs. Or, ce traitement des effluents a pour effet de produire principalement des boues difficilement filtrables. Toutefois, l'application du procédé de congélation à ces boues facilite leur filtration, après quoi les boues, qui contiennent encore 60 % d'eau environ sont directement incorporées dans le bitume fondu.

Au cours de cette période d'exploitation relativement longue, des modifications ont été régulièrement apportées aux installations d'origine, ce qui a contribué à l'amélioration du procédé de traitement et également à la sécurité de l'installation.

2. DESCRIPTION DU PROCEDE

Ce procédé "à haute température" ainsi qu'il est nommé depuis des

.../...

années comporte deux étapes bien distinctes :

1° la préparation de la boue à incorporer

: cette préparation est surtout nécessaire pour extraire dans toute la mesure du possible l'eau contenue dans la matière à incorporer, étant donné que la capacité d'évaporation de l'installation d'insolubilisation est limitée et qu'elle dépend en outre d'une série de facteurs qui sont à la base d'une installation bien conçue, à savoir la quantité de bitume, les dimensions de la cuve de mélange, la capacité calorifique des éléments chauffants par unité de surface etc.

Plus les boues sont sèches, plus élevée sera la capacité de traitement de l'installation. Des boues trop sèches présentent toutefois l'inconvénient de former de gros noyaux de boue durcie.

: la préparation des boues comprend essentiellement :

- une décantation des boues liquides par centrifugation ou par décantation jusqu'à que la concentration atteigne 7 %;
- une filtration des boues décantées, qui ont préalablement été congelées pendant un certain temps;
- une filtration sur filtre à vide jusqu'à l'obtention d'une matière solide contenant environ 60 % d'eau, ce qui permet leur passage dans l'installation d'insolubilisation.

2° l'insolubilisation proprement dite

L'homogénéité du mélange boues-bitume fait l'objet de la plus grande attention. Cette méthode de mélange possède les caractéristiques d'un encapsulage monomoléculaire, obtenu à l'aide d'une agitation intense par mélangeur rapide. Etant donné que le bitume ne peut pas être trop visqueux, il est nécessaire de porter le bain de bitume à une température élevée (240° C), ce qui contribue à améliorer sensiblement la capacité d'évaporation du dispositif à hauteur du tourbillon créé dans le bain. La vapeur d'eau qui se dégage avec force du bain de bitume fait mousser le bain. La formation de cette mousse peut être controlée par addition d'huile de silicone.

Un thermocouple disposé dans le bain de bitume maintient la température entre 215 et 240° C. Lorsque la température du bain baisse jusqu'au minimum, l'alimentation en boues est arrêtée tandis que les éléments chauffants sont remis en fonctionnement. Lorsque la

.../...

température atteint le maximum, l'inverse se produit, c'est à dire que le système de chauffage est mis hors circuit et l'alimentation en boue se poursuit.

3. DESCRIPTION DE L'INSTALLATION ACTUELLE

L'installation comprend :

1° une unité d'entreposage et de transfert des boues depuis le réservoir de collecte jusqu'à une centrifugeuse.

2° un réservoir de stockage de la boue ainsi épaissie, prévu pour l'alimenter.

3° l'installation de congélation, qui se compose de deux bacs d'un m^3 de contenance, bien isolés et dans lesquels sont fixées des plaques de congélation (et de réchauffage). Le froid, ou la chaleur, sont obtenus à l'aide d'une circulation de glycol (- 15° C ou + 30° C).

4° un réservoir destiné à alimenter le filtre à vide.

5° un filtre à vide (à tambour) pourvu d'un système de raclage automatique débouchant sur un tourillon d'alimentation.

6° un insolubilisateur qui assure le mélange des boues filtrées et du bitume.

Cette partie de l'installation comprend essentiellement :

6.1. une cuve de mélange, en acier ordinaire, chauffée électriquement par éléments immergés aussi bien que par manteau chauffant dans les parois. Ce manteau chauffant est divisé en deux zones; la zone inférieure possède un pouvoir calorifique de 15 KW tandis que celui de la zone supérieure est de 10 KW. Les 60 éléments immergés sont disposés suivant une géométrie bien déterminée et représentent un pouvoir calorifique global de 60 KW ce qui porte à 85 KW le pouvoir calorifique de l'insolubilisateur.

6.2. un mélangeur du type "VARICINETIC" (de 1.500 à 3.000 t/min.) équipé de 8 ailettes réglables en vue de maintenir à un niveau constant le processus de mélange dans le bain de bitume au fur et à mesure que la viscosité du mélange augmente en fonction de la teneur en boues.

.../...

6.3. un système d'alimentation à vis d'Archimède dont la vitesse de rotation est variable.

6.4. une vanne à boisseau sphérique, chauffée électriquement, pour la décharge du mélange.

6.5. un espace ventilé, avec protection, équipé d'un carrousel sur lequel sont disposés les récipients qui seront remplis du mélange boues-bitume traité.

6.6. un ensemble groupant les dispositifs d'extraction, d'épuration des gaz condensables ou non-condensables.

La vapeur d'eau qui se dégage du bain passe d'abord par un dévésiculeur fortement chauffé, qui retient les particules de bitume entraînées par la vapeur. Celle-ci passe ensuite par deux condenseurs à mélange, montés en série, où s'opère un lavage à l'eau froide provoquant la condensation.

Après condensation, les eaux sont recueillies dans un réservoir de collecte et dirigées vers le traitement des effluents radioactifs.

L'activité moyenne des gaz résiduels non condensables se répartit comme suit :

10^{-11} µci/ml alpha

10^{-9} µci/ml bêta-gamma

D'autre part, l'activité des solutions de lavage (débit 800 l./h) atteint les moyennes suivantes :

10^{-7} µci/ml alpha

5.10^{-5} µci/ml bêta-gamma.

L'entretien de routine qui s'effectue tous les 6 mois est toujours précédé d'un rinçage à l'aide de bitume frais.

Cet entretien est particulièrement indispensable en vue de remplacement des ailettes du mixer dont l'usure est provoquées par l'action abrasive de certaines espèces de boue ($BaSO_4$ par ex.) et en raison du colmatage progressif des éléments chauffants par une fraction calcinée du mélange bitume/boues.

Il faut enfin signaler que la vanne de décharge du mélange ne s'est obstruée qu'à de rares reprises.

.../...

7. EXPERIENCE ACQUISE

Depuis le milieu de l'année 1964 fonctionne au département de traitement des résidus radioactifs du C.E.N./S.C.K. à Mol une installation "MUMMIE" principalement conçue en vue de l'incorporation dans le bitume de concentrats produits par le traitement d'effluents liquides faiblement actifs.

Ces concentrats se composent essentiellement de boues formées de $Ca_3(PO_4)_2$ - $Al(OH)_3$ - $Mn_2Fe(CN)_6$ - $CuFe(CN)_6$ etc.

Le traitement s'effectue à raison de 16 heures/jour et a porté jusqu'à présent sur un total de $\pm$ 750 tonnes de boue humide (60 % d'eau en moyenne) soit environ 300 tonnes de boue sèche. Le mélange final de boue et de bitume ainsi obtenu représente environ 700 tonnes.

La boue insolubilisée provenait du traitement de $\pm$ 1.400.000 m^3 d'effluents liquides dont la radioactivité a été confinée dans un volume ultime de $\pm$ 550 m^3 de mélange boues/bitume (activité totale 10.000 Curies).

L'installation d'insolubilisation fonctionne en moyenne 4.000 heures par an avec un coefficient d'occupation d'environ 90 %.

Les caractéristiques du mélange boue/bitume sont les suivantes :

- densité : 1,3
- Ring & Ball : 110 à 120° C
- vitesse de lixiviation : 10^{-6} à 10^{-7} g/cm^2/jour (après 90 jours).

La concentration de particules huileuses présentes dans les vapeurs est de l'ordre de 500 mg/m^3 avant le circuit de refroidissement.

8. PRIX DE REVIENT

La capacité d'évaporation de l'installation d'insolubilisation à Mol est d'environ 80 à 60 l/heure, ce qui correspond à une quantité de 15 à 20 kg de boue sèche par heure (soit 65 à 70 kg de boue humide).

Ceci représente, dans le cas de l'installation d'insolubilisation en question, une production de 0,5 m^3 de mélange final boue/bitume par 1.000 m^3 d'effluents faiblement actifs traités (max. 10^{-2} $\mu Ci/ml$) pour un facteur de réduction de l'ordre de 2.000.

.../...

Les dépenses se répartissent comme suit :

Etape du traitement	Personnel	Energie	Produits Fournit.	TOTAL
- Epaississement des boues par centrifugation	380	30	-	410
- Congélation des boues	1.900	5	-	1.905
- Filtration	2.128	20	1.052	3.200
- Insolubilisation dans le bitume	2.280	575	1.950	4.805
- Traitement des distillats	152	-	2.568	2.720
- Décharge et conditionnement dans les récipients de stockage	760	10	940	1.710
TOTAL (FB)	7.600	640	6.510	14.750

9. DEVELOPPEMENT

Dans le cadre de la participation belge au programme, SNR la BELGONUCLEAIRE a entrepris la réalisation et la mise au point d'une unité pilote de traitement de déchets radioactifs liquides et de leur insolubilisation dans le bitume.

L'unité réalise un séchage préalable des solutions alimentées suivi d'un enrobage des résidus séchés dans le bitume.
Conçue comme une installation composée de matériels industriels, cette unité devait permettre la démonstration du procédé, d'abord avec des effluents type "réacteur à sodium", ensuite avec des solutions simulant les effluents produits par les LWR.

Le procédé comporte deux étapes principales :

1) un séchage préalable des solutions alimentées de manière à séparer les résidus secs de l'eau.

.../...

2) une incorporation du produit séché dans le bitume chaud.

9.1. Description résumée du procédé.

Commençant par l'alimentation en fluide à traiter, les principales étapes opérationnelles sont :

1) L'effluent prétraité dans un réservoir d'alimentation est pompé dans un sécheur rotatif chauffé à la vapeur où l'eau contenue dans la solution est évaporée. Les résidus tombent alors dans une trémie.
2) Les résidus séchés sont progressivement introduits dans un réservoirs de mélange contenant du bitume préchauffé. L'humidité résiduelle est éliminée par évaporation. Lorsque la quantité voulue de produits solides a été introduite, le mélange est déversé dans un fût normalisé en acier.
3) Les vapeurs dégagées passent par un séparateur ou un sécheur avant condensation. A partir du réservoir de condensat, une pompe rejette l'eau vers un filtre à cartouches.
4) Les gaz incondensables sont aspirés à travers une batterie de filtres avant d'être évacués. L'enceinte est ainsi maintenue sous pression.

La capacité d'une telle unité peut facilement varier de 100 à 200 l/h d'eau évaporée pour une exploitation continue pendant une période de temps très longue.

Le concept décrit ci-dessus est rapidement adaptable à l'usage de résines polymères ou de tout autre matériau de matrice utilisé comme agent de solidification.

9.2. Essais.

Des essais ont été effectués avec des solutions simulées d'effluents liquides radioactifs, soit des solutions chimiques, soit des solutions chimiques avec traceur radioactif. C'est ainsi que des solutions de sulfates, de nitrates, de phosphates, de borates, de détergents et des suspensions de résines en poudre ont été testées.
Quelques essais moins systématiques avec des boues de traitement d'effluents liquides et des résines en grains ont été également entrepris.

.../...

9.3. Conclusions.

Ces campagnes d'essais ont permis de tirer de nombreux enseignements.

Elles ont confirmé l'intérêt essentiel du "prétraitement" des solutions de manière à favoriser l'étape du séchage et à obtenir ensuite un produit bituminisé satisfaisant.

Le mode d'alimentation au sécheur ou éventuellement le type même du sécheur dépend également de la nature et de la composition des solutions à traiter.

Dans l'ensemble la technique du séchage et de l'incorporation dans le bitume des résidus séchés peut être maitrisée sans problème majeur moyennant une adaptation adéquate aux effluents en présence.

Citons encore deux avantages appréciables du procédé :

- chaque étape (c'est à dire traitement préalable, séchage, solidification) peut être réalisée et contrôlée indépendamment; ceci procure une souplesse certaine et permet un ajustement facile des paramètres en cas de changement des conditions de travail. Chaque étape peut même être réalisée sans interférence aucune avec les autres.
- une sortie séparée du matériau séché solide peut être prévue sur la trémie de stockage d'attente, évitant ainsi l'étape de solidification quand cette dernière n'est pas requise.

Une unité industrielle basée sur ces principes sera prochainement montée dans une centrale BWR en Suède.

Discussion

J. ORTEGA ABELLAN, Spain

Why do you use aluminium hydroxide [$Al(OH)_2$] instead of ferr hydroxide [$Fe(OH)_3$] for precipitation ? What are the advantages ?

N. VAN DE VOORDE, Belgium

Both products are used as scavengers to purify the water. $Fe(OH)_3$ has also a certain decontaminating effect, for example for ruthenium and cerium. In our new treatment concept, we aim principally at purification before ion exchange treatment.

J. ORTEGA ABELLAN, Spain

The problem of detergents is very important. Have you noted an influence of detergents on the fixation of fission products and particularly ^{90}Sr on $Fe(CN)_6Ni_2$ and CO_3Ca ?

N. VAN DE VOORDE, Belgium

Selective sorbents already exist which have satisfactory efficiency in a medium with a high detergent concentration.

N. FERNANDEZ, France

M. van de Voorde gave an operating cost of 3,000 BF/cubic metre of sludge before centrifugation. What it, at this stage, the solid material content of this sludge ?

N. VAN DE VOORDE, Belgium

At this stage, the solid material content is 7 %.

W. HILD, F.R. of Germany

As far as the ion exchange resins are concerned, we have noticed a degradation of the mixed bed ion exchangers too. The main degradation products were amines. We did not find any reaction between sulfates and bitumen as in your case. Incorporation of ion exchange resin can - as demonstrated by our investigation - be safely incorporated into bitumen without degradation at temperatures around 120°C.

J.E. STEWART, United States

During maintenance, what dose rates do personnel receive ?

N. VAN DE VOORDE, Belgium

Dose rates are normally below 50 mrem.

J.E. STEWART, United States

How often does maintenance occur ?

N. VAN DE VOORDE, Belgium

Normally every 6 months.

W. HILD, F.R. of Germany

I would like to comment on the question of detergents. We have faced the same problems at Karlsruhe, too. By imposing on the various effluent producers to cut down detergent consumption and to increase the price of effluent treatment, we arrived at a detergent consumption of about 100 kg for 3,500 kg laundry. The same is true for cutting down the consumption of decontamination agents. There we arrived at a drastic restriction of the consumption of chemicals, too.

Discussion

J. ORTEGA ABELLAN, Espagne

Pourquoi utilisez-vous l'hydroxyde d'aluminium [$Al(OH)_2$] au lieu de l'hydroxyde ferrique [$Fe(OH)_3$] pour la précipitation ? Quels en sont les avantages ?

N. VAN DE VOORDE, Belgique

Les deux produits sont employés comme produits de nettoyage (scavengers) pour clarifier l'eau. Le $Fe(OH)_3$ donne en plus un certain effet de décontamination, par exemple pour le ruthénium et le cérium. Dans notre nouvelle conception du traitement, on cherche surtout la clarification avant le traitement par échange d'ions.

J. ORTEGA ABELLAN, Espagne

Le problème des détergents est très important. Avez-vous constaté l'influence des détergents sur la fixation des produits de fission et notamment dans la fixation du ^{90}Sr sur le $Fe(CN)_6Ni_2$ et le CO_3Ca ?

N. VAN DE VOORDE, Belgique

Il existe déjà des sorbents sélectifs qui ont une bonne action dans un milieu à haute concentration de détergents.

N. FERNANDEZ, France

M. van de Voorde a donné comme coût d'exploitation 3.000 FB par m^3 de boue avant centrifugation. Quelle est la teneur de cette boue en matière solide, à ce stade ?

N. VAN DE VOORDE, Belgique

La teneur en matière solide, à ce stade, est de 7 %.

W. HILD, R.F. d'Allemagne

En ce qui concerne les résines échangeuses d'ions, nous avons observé une dégradation des échangeurs d'ions à lit mixte également, les principaux produits de dégradation étant les amines. Nous n'avons pas constaté de réaction entre les sulphates et le bitume comme dans votre cas. Les résines échangeuses d'ions peuvent, comme l'ont montré nos recherches, être incorporées en toute sûreté dans du bitume sans entraîner de dégradation à des températures avoisinant 120°C.

J.E. STEWART, Etats-Unis

Pendant les travaux d'entretien, quelles ont été les doses d'irradiation reçues par le personnel ?

N. VAN DE VOORDE, Belgique

Les doses sont normalement inférieures à 50 mrems.

J.E. STEWART, Etats-Unis

Quelle est la fréquence des travaux d'entretien ?

N. VAN DE VOORDE, Belgique

Normalement, tous les six mois.

W. HILD, R.F. d'Allemagne

Je souhaiterais faire une observation sur la question des détergents. Nous avons rencontré les mêmes problèmes à Karlsruhe. En obligeant les divers producteurs d'effluents à réduire la consommation de détergents et à augmenter le prix du traitement des effluents, nous sommes parvenus à une consommation d'une centaine de kg de détergent pour 3.500 kg de lessive. Il en va de même pour la consommation d'agents de décontamination. Dans ce cas, nous sommes parvenus à réduire aussi considérablement la consommation de produits chimiques.

CONDITIONNEMENT DANS LE BITUME DES DECHETS RADIOACTIFS DE FAIBLE ET MOYENNE ACTIVITE

G.LEFILLATRE

C.E.A.
Service de Chimie Appliquée
B.P. N° 1
13115 Saint Paul lez Durance - France

Résumé

La communication fait état des travaux de développement français portant principalement sur :

- le bitumage des résidus radioactifs produits par les centrales à eau légère,
- le bitumage direct sans préconcentration des effluents liquides
- l'expérimentation d'enfouissement dans le sol pendant 18 mois de blocs de 100 l de concentrats solidifiés par le bitume.

Elle mentionne également l'expérience acquise en France depuis 1971,

- à VALDUC et à SACLAY avec des stations équipées d'évaporateurs à couche mince,
- à MARCOULE pour le condiditionnement de trilaurylamine et de tributylphosphate.

Enfin elle décrit sommairement la future installation de CADARACHE équipée d'une extrudeuse à double vis, dont la mise en service est prévue en 1977.

Summary

The document présents french opérations concerning mainly :

- the bituminization of radioactive wastes produced by light water reactors,
- the direct bituminizing, without preconcentration, liquid effluents,
- the experimentation for 18 months with the land burial of evaporation concentrates solidified by bitumen into blocks of 100 liters.

This paper mentions also the knowledge that they have acquired in France :

- in VALDUC and in SACLAY with facilities equipped with thin film evaporators,
- in MARCOULE with the conditioning of trilaurylamine and tributylphosphate.

At last the future bituminizing facility in CADARACHE, equipped with twin screws extruder, designed starting in 1977, is succinctly described.

I - INTRODUCTION

Depuis seize ans le CEA a entrepris l'étude en laboratoire, la mise au point industrielle et l'exploitation du bitumage des effluents radioactifs de faible et moyenne activité.
Le choix des procédés mis en oeuvre a été constamment guidé par le souci de réduire le plus possible les risques d'exposition des travailleurs aux radiations lors des opérations de conditionnement des déchets, d'abaisser à un niveau très faible les risques chimiques liés à l'opération de bitumage et d'assurer un confinement durable et sûr des déchets radioactifs, tout en réduisant leur volume et leur poids.
Ces résultats ont été recherchés et obtenus par l'adoption de procédés continus faisant appel à des matériels éprouvés qui permettent notamment :

- un conditionnement des déchets sans intervention manuelle,
- un temps de chauffage et de séjour des produits très court
- un contrôle précis des températures
- la production d'un enrobé homogène.

La tendance actuelle au CEA est de remplacer progressivement la cimentation des résidus liquides par le bitumage.

II - TRAVAUX DE DEVELOPPEMENT RECENTS

Les études de développement entreprises depuis 1971, ont porté sur :

1 - le conditionnement de deux types d'effluents ou déchets répondant aux besoins propres des centres industriels du CEA :

- la solidification par le bitume des boues de coprécipitation chimique et des concentrats d'évaporation produits par la Station de Traitement des Effluents du Centre de SACLAY
- la solidification par le bitume des sels dissous dans les solutions concentrées produites par l'évaporation des effluents liquides du centre de CADARACHE.

2 - le conditionnement des deux types d'effluents ou déchets, répondant aux besoins des Sociétés d'électricité, produits par les centrales nucléaires de puissance à eau légère, d'une part ceux de la filière bouillante, d'autre part ceux de la filière pressurisée.

3 - le bitumage direct, sans préconcentration, des effluents à faible minéralisation (supérieure à 1 g/l) provenant aussi bien des centrales nucléaires de puissance que des centres d'études nucléaires. Ces études, initiées au préalable en laboratoire, ont été développées dans l'installation pilote du centre de CADARACHE, équipée d'un évaporateur à couche mince vertical LUWA L150, d'une capacité de traitement industrielle (50 à 60 l/h).

4 - l'incorporation de solvants organiques lourds: trilaurylamine et tributylphosphate, provenant du cycle de retraitement des combustibles irradiés, dans les boues de coprécipitation chimique solidifiées par le bitume.
La mise au point de ce type de conditionnement a été faite, après essais préliminaires en laboratoire, sur la chaîne d'enrobage de la Station de Traitement des Effluents du Centre de MARCOULE.

5 - les études de sûreté sur les déchets solidifiés par le bitume comprenant :

- la lixiviation de la radioactivité dans les eaux (eau déminéralisée - eau de nappe phréatique - eau de mer)
- la stabilité thermique par analyse thermique gravimétrique et différentielle, des déchets pendant l'opération de conditionnement et au stockage.
- la tenue biologique vis à vis des microorganismes dans le cas

d'un enfouissement des déchets dans le sol.
Ala suite de ces travaux de développement des installations industrielles ont été mises en service , ou le seront prochainement.

1 - Conditionnement des concentrats d'évaporation des Centres d'Etudes Nucléaires.

Deux Stations de Traitement d'Effluents équipées l'une et l'autre d'un évaporateur à couche mince LUWA L 150 sont actuellement en fonctionnement à VALDUC et à SACLAY. La première [1] mise en actif en 1971 a conditionné plus de 1200 fûts d'enrobé d'une activité spécifique allant de 0,1 à 10 Ci/m^3 α.
La seconde [2] plus récente, a solidifié les premiers concentrats d'évaporation actifs en juin 1975, et a depuis conditionné plus de 300 fûts de 200 l d'enrobé d'une activité spécifique allant de 1 à 20 Ci/m^3 α,β,γ.
Une troisième Station de Traitement d'Effluents, celle de CADARACHE, disposera en 1977 d'un poste de bitumage équipé d'une extrudeuse W.P. à double vis ZDS -T83, d'une capacité de solidification annuelle d'environ 260 m^3 de concentrats à 250 - 300 g/l, dont l'activité spécifique sera comprise entre 1 et 10 Ci/m^3 α,β,γ.
Cette nouvelle installation est décrite brièvement à la fin de cette communication.

2 - Conditionnement des résidus produits par les Centrales Nucléaires à eau légère.

2.1. Résidus des Centrales à eau bouillante

En ce qui concerne les résidus des Centrales type BWR, les études de mise au point du bitumage de ces déchets, faites pour le compte de SAINT GOBAIN TECHNIQUES NOUVELLES, ont servi à la conception, la construction et au démarrage de l'installation de BARSEBACK en Suède (2 BWR SYDKRAFT de 560 MWe) livrée par S.G.N. en Juillet 1974.
L'enrobage des résidus suivants a été réalisé avec l'évaporateur à couche mince :

- concentrats d'évaporateur, principalement des solutions concentrées à 200 g/l de sulfate de sodium issues des éluats des échangeurs d'ions,
- résines échangeuses d'ions, prébroyées ou non avant le bitumage, et résines Microionex,
- adjuvants et boues de filtration (diatomées - solka - floc ect...).

Ces différents types d'effluents peuvent être traités après mélange ou séparément.
A titre d'exemple, les enrobés issus du traitement de solidification de ces déchets par le bitume M 40/50 ont des volumes, rapportés au poids d'extrait sec, suivants :

- concentrats à 200 g/l de Na2 SO4 : 190 l d'enrobé pour 100 kg d'extrait sec (160°C pendant 5h). Teneur en extrait sec : normale 40 % - maximale 45 %
- résines AMBERLITE IRN 77 et 78 non broyées : 230 l d'enrobé pour 100 kg d'extrait sec. Teneur en extrait sec : normale 40 % maximale 50 %.
- Diatomées CLARCEL DIC 3:370 l d'enrobé pour 100 kg d'extrait sec. Teneur en extrait sec: normale 25 % - maximale 33 %
- mélange en volume 1/2 de concentrats à 200 g/l Na2 SO4 et 1/2 de résines IRN 77 et 78 ; 220 l d'enrobé pour 100 kg d'extrait sec. Teneur en extrait sec : normale 40 % - maximale 45 %,
- mélange en volume 1/3 de concentrats 200 g/l NA2 SO4, 1/3 de résines IRN 77 et 78, 1/3 diatomées CLARCEL DIC 3 : 225 l d'enrobé pour 100 kg d'extrait sec. Teneur en extrait sec : normale 36 % maximale 40 %.

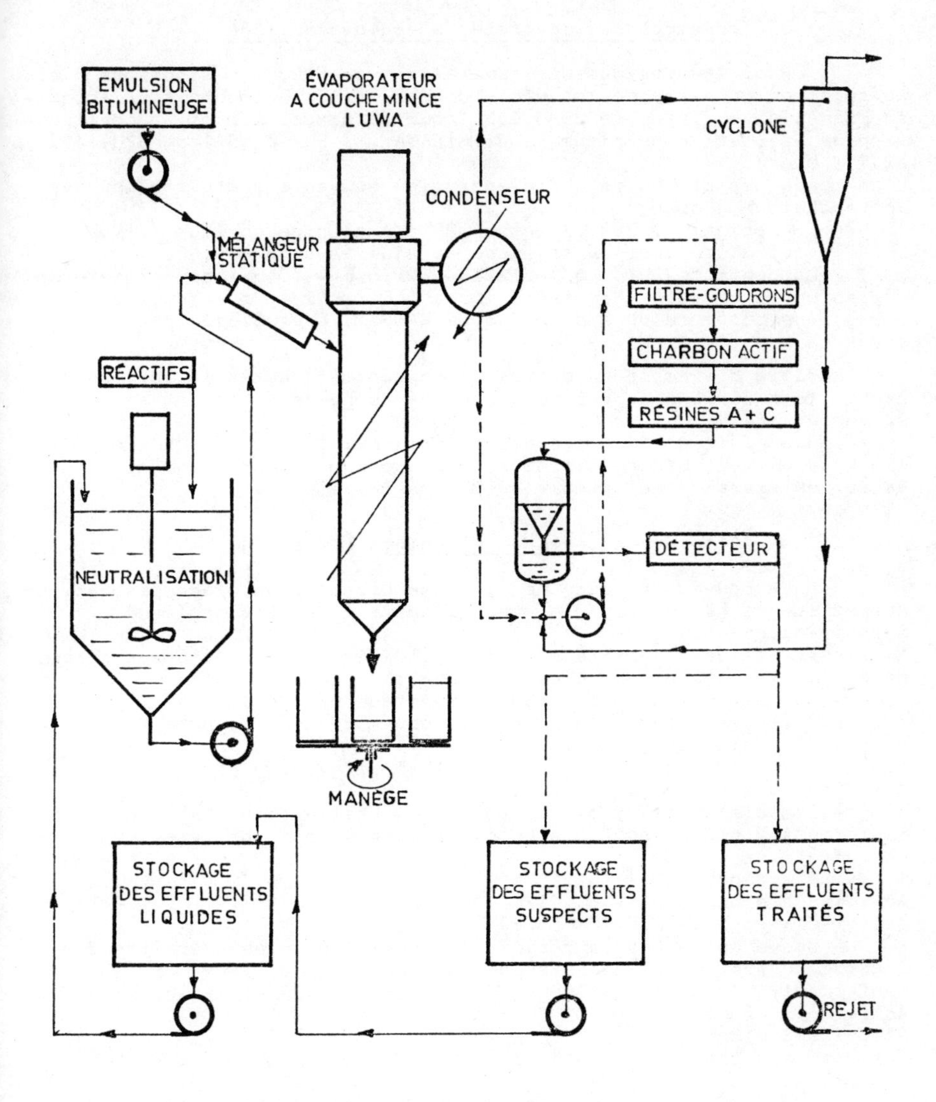

Fig:1_ BITUMAGE DIRECT DES EFFLUENTS RADIOACTIFS SANS PRÉCONCENTRATION.

2.2. Résidus des centrales à eau pressurisée

Pour les résidus des centrales type PWR, les études de mises au point de bitumage de ces déchets vont servir à la Société Japonaise EBARA, sous licenciée de S.G.N. pour la conception et la construction de la Station de bitumage de MIHAMA (3 PWR KANSAI de 320, 470 et 793 MWe).
L'enrobage des concentrats d'évaporation suivants a été réalisé avec l'évaporateur à couche mince.

- Concentrats à 200 g/l et à 300 g/l d'acide borique (type SENA) neutralisés par de la soude et ajustés à pH = 9,65.
- Concentrats à 132 g/l d'acide borique - 10 g/l de chromate de potassium 10 g/l d'oxydes métalliques (Fe - Ni - Cr - Mn - Co) 0,9 g/l de chlorure de sodium (type KANSAI) neutralisés par de la soude et ajustés à pH = 9,65.

A titre d'exemple un enrobé de concentrats borés solidifiés par le bitume M 40/50 a le volume, rapporté au poids d'extrait sec, suivant :

concentrats à 300 g/l d'acide borique : 190 l d'enrobé pour 100 kg d'extrait sec de tétraborate de sodium (160°C pendant 5 h).
Teneur en extrait sec normale 40 % - maximale 50 %.

3 - Bitumage direct des effluents [fig.1]

Ce nouveau procédé de bitumage direct sans préconcentration des effluents [2] et [3] n'a, pour l'instant pas été appliqué industriellement.
Plusieurs effluents actifs ont été solidifiés par bitumage direct dans l'installation pilote de CADARACHE.

- Des effluents usés issus du réacteur PWR de la SENA (300 MWe) pH = 9,2 renfermant 2,02 g/l de sels dont 320 mg/l de bore.
Activité γ globale $1{,}3\ 10^{-2}$ Ci/m^3
Activité β globale 1,26 Ci/m^3 dont 1,25 Ci/m^3 de 3 H
Activité α globale 8,5 10^{-5}Ci/m^3
- Des effluents usuels provenant du Centre de SACLAY, pH = 6,6, renfermant 2 g/l de sels à prédominance phosphate, nitrate, chlorure et sulfate de sodium
Activité γ globale 2,4 10^{-2}Ci/m^3
Activité β globale 8,6 10^{-2}Ci/m^3
Activité α globale 6,6 10^{-4}Ci/m^3
- Des suspensions de résines AMBERLITE saturées extraites de l'unité de traitement des piscines de dégainage de LA HAGUE renfermant 10 g/l d'un mélange de résines IRA 400 et IR 120 pH = 7,8
Activité γ globale 2,25 10^{-1}Ci/m^3
Activité β globale 5,5 10^{-4}Ci/m^3
Activité α globale 4 10^{-4}Ci/m^3

3.1. Facteurs de décontamination de l'installation pilote

Les facteurs de décontamination obtenus à la suite des différents postes de traitement de l'installation sont mentionnés dans le tableau I.

TABLEAU I

F.D.	Total	^{137}Cs	^{54}Mn	^{75}Se	^{134}Cs	^{106}Ru	^{58}Co	^{60}Co	^{131}I	^{90}Sr
Evaporateur à couche mince LUWA 150	860 à 1200	800 à 1500	350 à 1400	80 à 300	900 à 1900	170 à 330	150 à 360	160 à 390	100 à 160	3000 à 30000
Filtration sur charbon actif	900 à 3000	1100 à 2500	800 à 1900	500 à 1500	1300 à 2400	190 à 350	—	175 à 400	—	30000 à 60000
Echange ionique	1500 à 6000	1800 à 6000	1500 à 5500	600 à 1700	200 à 10000	370 à 1500	—	450 à 2000	—	100000 à 300000

3.2. Caractéristiques des distillats.

Avant traitement :
- conductivité 3 à 5 10^{-5} mho/cm
  pH = 8,3 à 9,4 pour les effluents
  pH = 7,4 à 8,2 pour les résines
  matières organiques (huiles et goudrons): 6 à 50 mg/l

Après traitement :
- pH = 6,8 - 7,5
  matières organiques (huiles et goudrons) : 1,5 à 4 mg/l
  activité spécifique ; toujours inférieure à la CMAP (hormis le tritium).

3.3. Caractéristiques des enrobés.

Densité réelle 1,38 à 1,57 pour les effluents - 1,23 pour les résines
Densité apparente 1,25 à 1,48 pour les effluents - 1,23 pour les résines
Teneur en eau 0 à 0,7 %
Teneur en bitume 60 à 50 %
Poids de ramollissement 73 à 78 °C

4 - Conditionnement des solvants organiques lourds

L'évaporation des solvants organiques lourds est réalisée industriellement dans la station de bitumage du Centre de MARCOULE qui utilise le procédé par effluence.[4]

Dans une première extrudeuse ZSK 120/3500, les boues chimiques sont enrobées par du bitume M 40/50 ou 80/100 chaud à l'aide d'émulgateurs. L'émulsion bitumineuse instable est rompue, ce qui entraîne la séparation mécanique d'environ 90 % de l'eau de constitution des boues. L'enrobé sortant de ce premier étage contient encore de 7 à 10 % d'eau. Il est introduit sous pression dans une deuxième extrudeuse ZHS 250/3300 où il est complètement deshydraté avant d'être coulé dans un fût de 225 l. Les solvants organiques lourds tels que la trilaurylamine et le tributylphosphate, dilués dans un solvant léger (dodécane - tertiobutylbenzène), sont injectés en continu à l'entrée de la deuxième extrudeuse de séchage grâce à une pompe doseuse à engrenage (20-80 l/h).

Les solvants contaminés sont approvisionnés par camion citerne, d'où ils sont transférés pneumatiquement dans une cuve de stockage de 2 m^3 ventilée. Toute l'installation d'alimentation et de conditionnement est réalisée en matériel anti-déflagrant.

4.1. Conditionnement de trilaurylamine T.L.A.

Le mélange trilaurylamine 15 % - Solgyl 54 B 85 % en volume est introduit à 60 l/h pour un débit de traitement de 400 l/h de boues à base de phosphate de fer, d'hydroxydes de fer et de cuivre et de ferrocyanure de nickel.
La température maximale de la machine ZHS ne dépasse pas 170°C
L'enrobé (d=1,35) qui coule à 158°C renferme 52,4 % de bitume M 80/100, et 4,[illegible] % de trilaurylamine, sans aucune trace de solvant (Pt Eb = 162 - 172 °C). Son point de ramollissement est supérieur à celui d'un enrobé sans trilaurylamine à la même teneur en bitume : 158°C au lieu de 135 °C. La viscosité de l'enrobé à la sortie de la ZHS est d'ailleurs nettement augmentée.

4.2. Conditionnement de tributylphosphate T.B.P.

Le mélange tributylphosphate 30 % dodécane 70 % en volume (activités spécifique β, γ, 1,8 Ci/m^3 à 97 % de 106 Ru et α 6,3 10^{-3} Ci/m^3) est introduit à 25 l/h pour un débit de traitement de 500l/h de boues de coprécipitation chimique.
La température maximale de la machine ZHS ne dépasse pas 160°C. Contrairement à la trilaurylamine, le tributylphosphaste diminue sensiblement la viscosité de l'enrobé et sa teneur dans l'enrobé ne doit pas dépasser 3 % pour éviter un point de ramollissement inférieur à 100°C.

5 - Etudes de sûreté.

5.1. Lixiviation de la radioactivité dans l'eau.

Tous les déchets solidifiés par le bitume élaborés au cours des études et fabriqués industriellement sont testés suivant la norme AIEA.
L'expérience de lixiviation dans l'eau ordinaire portant sur des blocs réels de 150l à 200 l de boues solidifiées par le bitume est poursuivie à MARCOULE [5]
Au bout de 1800 jours le taux de lixiation $\frac{a}{A}$ est de 8.10^{-4} dans l'eau ordinaire, en tenant compte de la décroissance, et la vitesse de lixiviation de 2,5 10^{-6}cm/jour.
Les tests de lixiviation en laboratoire sont effectués à CADARACHE, et pour l'ensemble des déchets : concentrats d'évaporation, effluents divers, résines, boues de coprécipitation chimique avec ou sans TBP ou TLA, leurs enrobés ont en moyenne les vitesses de liviation suivantes dans l'eau déminéralisée, exprimées en cm/jour.
1.10^{-6} à 3.10^{-8} pour le 137 Cs
1.10^{-5} à 1.10^{-7} pour le 90 Sr
1.10^{-5} à 1.10^{-6} pour le 106 Ru
1.10^{-5} à 1.10^{-6} pour le 60 Co
1.10^{-8} à 3.10^{-9} pour les émetteurs (238 Pu-239Pu-241Am)

5.2. Stabilité thermique.

Une analyse thermique gravimétrique et différentielle est pratiquée systématiquement sur chaque type de déchet susceptible d'être conditionné dans le bitume.
Cette analyse permet de déceler les réactions exothermiques brutales pouvant provoquer des risques d'explosion, de mettre en évidence les températures d'inflammation spontanée des différents déchets et de se rendre compte de la stabilité thermique des enrobés à différentes températures de stockage (250° à 400 °C) pendant un temps donné.
Dans le cas des concentrats d'évaporation, en particulier ceux des Centres d'Etudes Nucléaires, il est intéressant de connaitre l'influence de certains composés particulièrement oxydants ou thermolabiles sur la stabilité thermique des déchets à conditionner

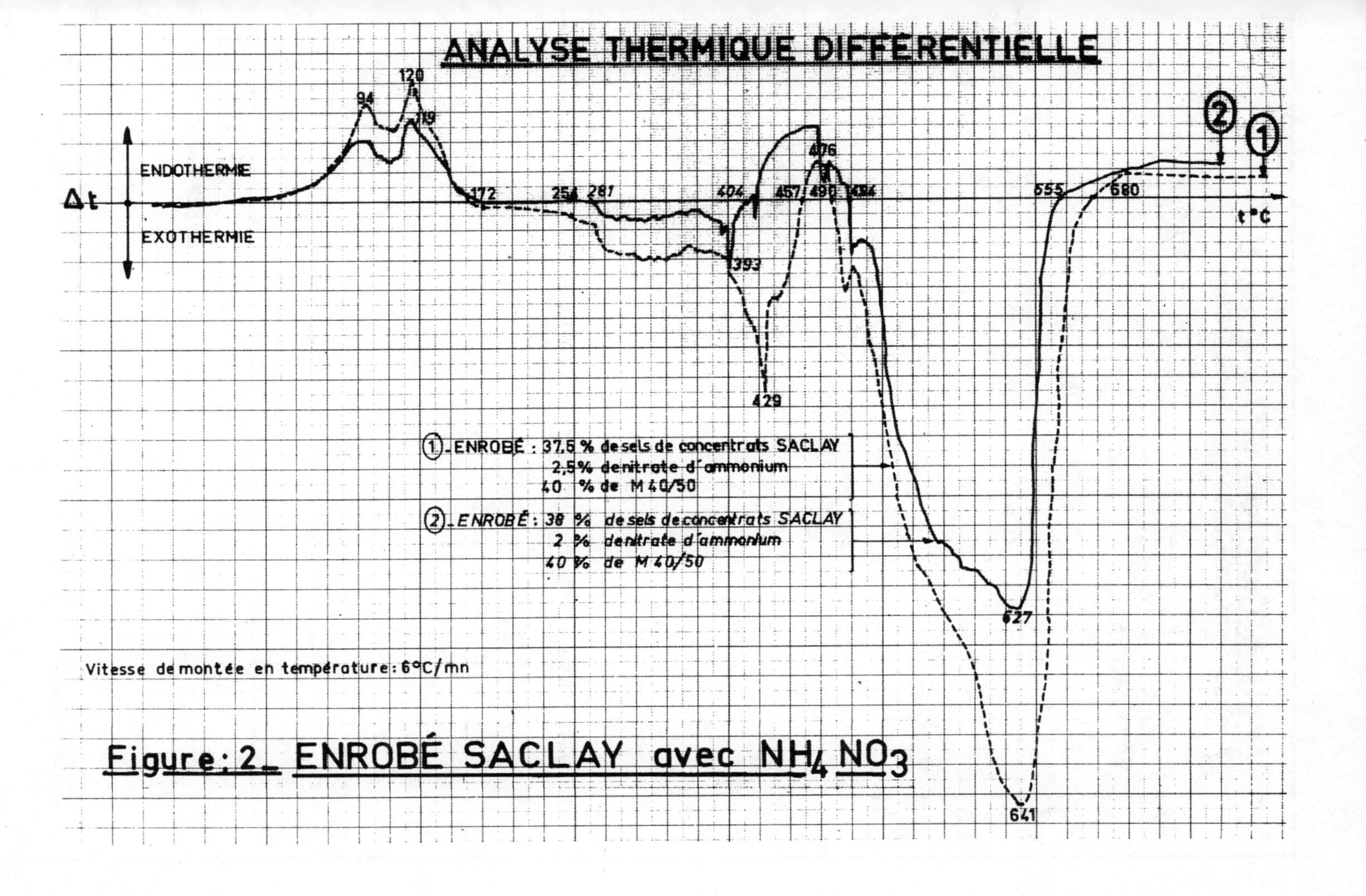

Figure : 2_ ENROBÉ SACLAY avec NH_4NO_3

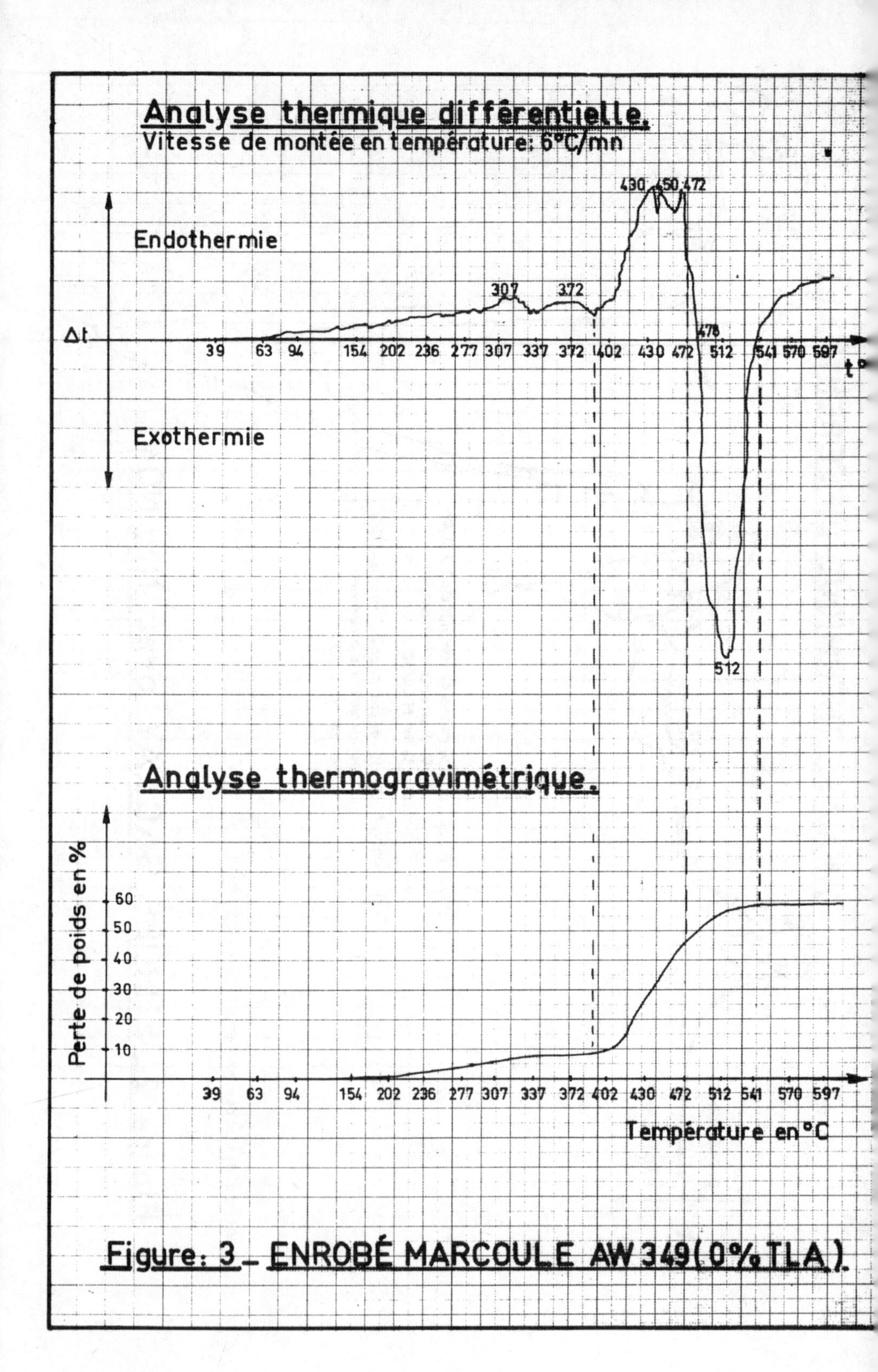

Figure: 3 _ ENROBÉ MARCOULE AW 349 (0% TLA).

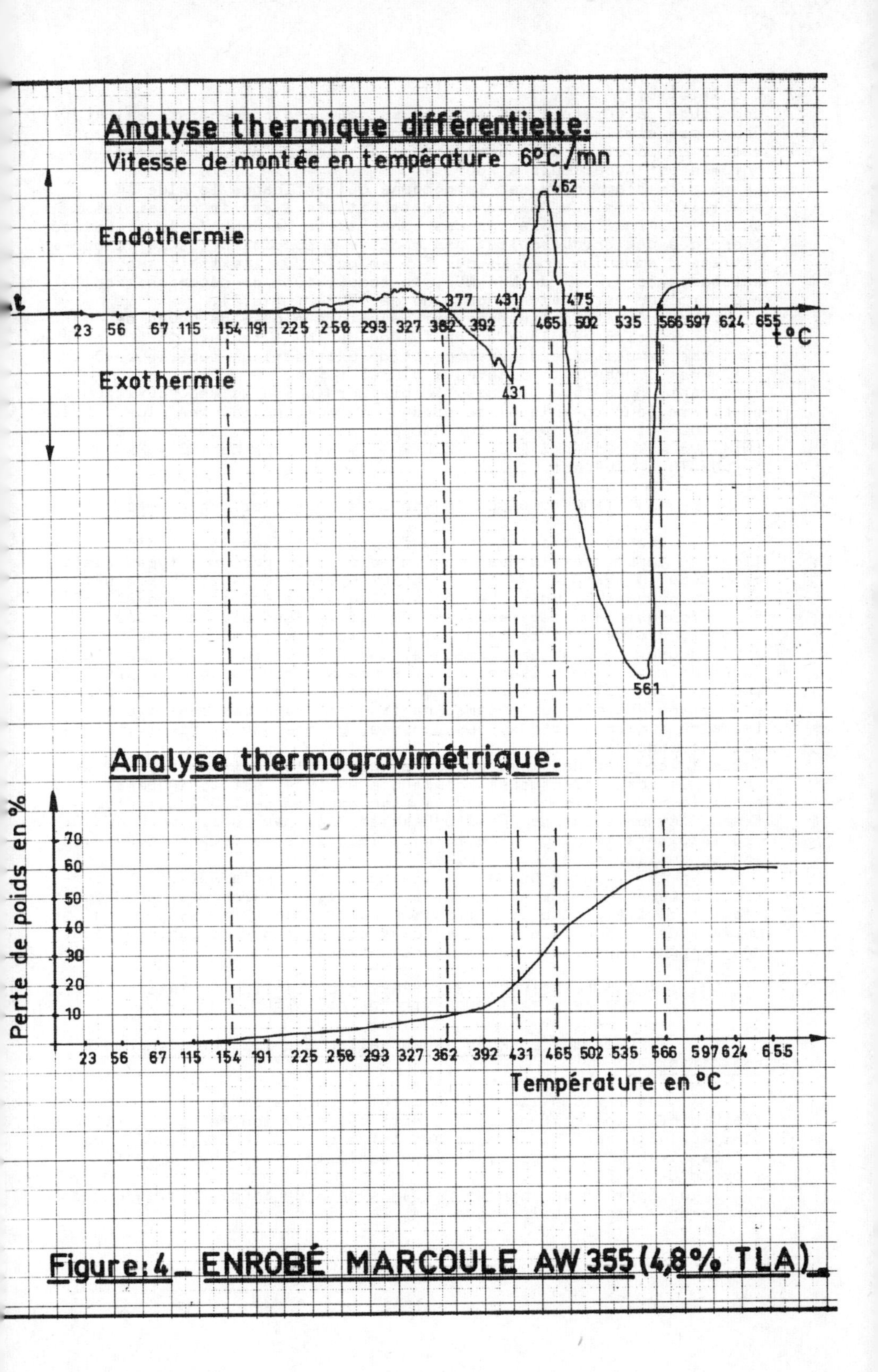

Figure: 4_ ENROBÉ MARCOULE AW 355 (4,8% TLA)_

par le bitume.
Une étude de sûreté a été entreprise pour les concentrats de SACLAY, une autre similaire est en cours pour ceux de CADARACHE. La température de début de décomposition de l'ensemble des concentrats bitumés est supérieure à 280°C.
Le nitrate d'ammonium est le composé qui influence le plus la stabilité thermique des concentrats bitumés à partir de faibles teneurs . Ainsi dans le cas des concentrats de SACLAY, la stabilité thermique des enrobés n'est pas modifiée à condition que la teneur en nitrate d'ammonium n'excède pas 5 % en poids de l'extrait sec à enrober; avec 6,25 % de NH_4 NO_3, la température de décomposition exothermique est abaissée de plus de 100°C (fig.2).

Les déchets solidifiés par le bitume à base de boues de coprécipitation chimique et de TLA ou de TBP (teneurs $\leq$5 %) présentent une température de décomposition exothermique supérieure à celle des concentrats bitumés : 310°C. (fig.3 et 4).
Cependant, dans tous les cas, la première phase de décomposition du bitume (endothermique) se produit à des températures plus élevées 470°- 480° et n'est pas influencée par celle des autres composés (NH_4 NO_3 - $KClO_4$ - K_2 CrO_4 - K_2Cr_2 O_7 - K ClO_3 - Na NO_3 - KNO_3 - Na NO_2-TBP - TLA ect...).

5.3. Tenue biologique vis à vis des microorganismes.

Une expérimentation d'enfouissement dans le sol d'échantillons de concentrats d'évaporation solidifiés par trois types distincts de bitume (M 40/50 - H 80/90 et R 90/40) a été poursuivie à partir de 1973 pendant 31 mois. Ces échantillons ont été enterrés à 0,50 m de profondeur dans deux sols différents ; d'une part dans une prairie, drainée naturellement ; d'autre part dans une zone marécageuse, constamment humide et inondée pendant 5 mois de l'année.
Après 31 mois d'enfouissement les examens microbiologiques pratiqués sur les échantillons entérrés ne permettent pas de mettre en évidence une action des microorganismes se trouvant dans ces deux types de terrains (germes totaux, pseudomonas, germes dénitrifiants, germes sulfito-réducteurs).
Actuellement une deuxième expérimentation d'enfouissement dans le sol de la zone marécageuse, commencée en 1974, est toujours en cours après 18 mois. Elle porte sur un bloc inactif de 100 l de concentrats d'évaporation de CADARACHE solidifiés par du bitume M 40/50.
Les résultats sont, pour le moment, identiques à ceux obtenus avec les échantillons de laboratoire.

En 1973, une étude entreprise pendant 8 mois sur des concentrats bitumés de même nature dans des milieux aérobies et anaérobies n'avait révélé aucune prolifération particulière de germes à la surface des échantillons placés dans des colonnes de sable où était développée une culture bactérienne intense.

III - INSTALLATION DE BITUMAGE DE LA STATION DE TRAITEMENT DES EFFLUENTS DU CENTRE DE CADARACHE

1. Caractéristiques essentielles des résidus liquides à solidifier

Les résidus liquides à bitumer correspondent à des concentrats d'évaporation dont la salinité est comprise entre 250 et 300g/l.
Leur activité spécifique β, γ est inférieure à 10 Ci/m^3 (80 % de 137 Cs et 10 % de 60 Co), et leur activité spécifique α est de l'ordre de 1.10^{-3}Ci/m^3.

La répartition moyenne des sels dans un concentrat, avant le traitement d'insolubilisation, est la suivante :

$NaNO_3$	67,7 %	NH_4NO_3	10,5 %
$Na_3PO_4,12H_2O$	9,7 %	$Na_2B_4O_7,10H_2O$	2,9 %
Na_2SO_4	2,7 %	Na_2CO_3	2,5 %
$NaCl$	1,6 %	NaF	1,1 %
$CaCl_2,2H_2O$	0,8 %	KCl	0,5 %

Le traitement de coprécipitation consiste à ajuster le pH des concentrats à 8,5 et à ajouter 2,8 g/l de ferrocyanure de potassium et 2,9 g/l de sulfate de nickel pour insolubiliser le 137 Cs par formation de précipité de ferrocyanure de nickel '' in situ ''.

2 - Description sommaire de l'installation de bitumage

L'ensemble du poste de bitumage, installé dans le sous-sol du bâtiment abritant l'évaporateur primaire, à côté du poste de cimentation, comprend :

2.1. Un poste d'alimentation du bitume comprenant une cuve de stockage de bitume fondu calorifugée de 30 m^3, située à l'extérieur du bâtiment, et réchauffée par de la vapeur sous 3 bars.
Une pompe de circulation distribue par circuit bouclé du bitume à une pompe doseuse à engrenage alimentant en continu la machine d'enrobage.

2.2. Un poste de préparation et de stockage des concentrats disposant de deux cuves en acier inoxydable de 3 m^3 équipées d'agitateurs et placées à l'intérieur d'une protection biologique de 30 cm de béton. La première conçue pour traiter en discontinu des lots d'environ 2,3 m^3 de concentrats provenant de la vidange de l'évaporateur primaire, la deuxième prévue pour stocker les concentrats traités et alimenter en continu la machine d'enrobage à l'aide d'une roue doseuse couplée avec une pompe centrifuge auto - amorçante.

2.3. Un poste d'enrobage comprenant une extrudeuse ZDS-T83 à double vis et un poste de remplissage de fûts, positionnés dans un caisson étanche ventilé en acier inoxydable de 5m x2,6m x2,6m H, protégé par un mur de 10 cm de plomb. Seule la partie opératoire active de la machine de 2,8m est placée à l'intérieur dans l'axe longitudinal du caisson.
La partie mécanique d'entraînement d'une longueur identique est montée à l'extérieur. Sur le toit de l'enceinte sont disposés quatre hublots permettant de voir l'intérieur des cheminées de la machine.
A la sortie de l'extrudeuse, chauffée à la vapeur sous 8 bars, l'enrobé coule dans un fût de 225 l positionné sur un plateau tournant à 3 postes de réception.
L'opération de remplissage est contrôlée par un détecteur de niveau, type bulle à bulle, et visuellement grâce à deux hublots latéraux.
Les fûts après un refroidissement d'environ 10 h. sont évacués, hors du caisson, au moyen d'un transporteur à rouleaux après ouverture des deux portes coulissantes α et γ. Ces fûts transitent dans un sas protégé où ils sont pesés, puis mis dans des cloches de protection de 9,5 cm d'acier avant d'être évacués vers l'aire de stockage temporaire des déchets du site de CADARACHE.

2.4. Un poste de filtration en continu des distillats chargés en huiles et goudrons (50 à 100 mg/l) provement des quatre condenseurs placés à la sortie des cheminées de l'extrudeuse. Ce poste est constitué par une pompe centrifuge de distribution, un filtre à 96 cartouches perdues chargées en Sorbocel et un circuit de contrôle de débit et de perte de charge. Les distillats épurés sont envoyés dans une cuve de stockage de 5 m^3.
. Suivant leur activité spécifique, ces effluents sont recyclés en tête d'évaporateur primaire ou rejetés.

3 - Résultats d'exploitation escomptés.

Cette installation est conçue pour conditionner dans le bitume 55 l/h de concentrats à 250 g/l de sels.

Les opérations s'effectueront par postes continus de 8 h depuis le lundi matin jusqu'au vendredi après midi, ce qui permettra de traiter environ 5,5 m^3 de concentrats par semaine.

Le facteur de concentration de volume des résidus liquides après solidification sera légèrement supérieur à 3.

Le facteur de décontamination global du poste d'enrobage atteindra 10^4.

IV - CONCLUSION

Le conditionnement dans le bitume des déchets radioactifs de faible et moyenne activité a maintenant atteint le stade industriel ; en France et dans le monde des installations de bitumage fonctionnent de façon satisfaisante.
Certes, le bitumage présente certains risques liés à la nature même du matériau d'enrobage : inflammabilité possible, le bitume est une matière organique - fluage prévisible des déchets lors de la destruction par corrosion des conteneurs métalliques au cours du stockage, le bitume est une matière plastique.
Cependant, même si ce conditionnement n'est pas parfait, il permet d'assurer, avec un dégré de sûreté convenable et à un coût raisonnable, le stockage des déchets radioactifs de faible et moyenne activité.

REFERENCES BIBLIOGRAPHIQUES

[1] P.POTTIER - R.ANDRIOT - D.CUAZ
Progrès dans les techniques de traitement et de conditionnement des effluents dans les centres de recherche - 4ème conférence internationale sur l'utilisation pacifique de l'Energie Atomique GENEVE Septembre 1971.
AIEA -A/conf. 49/P/621 -pp 332 -334 VIENNE 1972.

[2] G.LEFILLATRE-J.LECONNETABLE
Bitumage des résidus radioactifs - Problèmes de sûreté et domaines d'application.
Conférence nucléaire européenne.
La maturité de l'énergie nucléaire - PARIS Avril 1975

[3] G.LEFILLATRE
Progrés dans les techniques de bitumage des effluents liquides des centrales nucléaires à eau pressurisée.
Colloque international sur la gestion des déchets radioactifs provenant du cycle du combustible nucléaire - VIENNE Mars 1976.

[4] N. FERNANDEZ
Enrobage bitumeux des boues de traitement des effluents radioactifs - Réalisation industrielle - Energie Nucléaire - Vol. 11 n° 6, pp 357 - 365 (1969).

[5] J.RODIER - G.LEFILLATRE
Etude de la diffusion de la radioactivité de blocs d'enrobés bitumeux en provenance d'un atelier d'enrobage industriel.
Rapport CEA R.3743(1969).

Discussion

K. BRODERSEN, Denmark

I understant that you have made some leaching tests on full scale bitumen blocks ; is the leaching rate obtained from such experiments the same or comparable with leaching rates measured using the standard lab-scale IAEA test ?

G. LEFILLATRE, France

Yes, the results for leaching tests with full scale bitumen concentrates are equal and even better than those obtained with lab-scale samples [cf. CEA R3743 report (1969)].

W. HILD, F.R. of Germany

At Valduc, do you incorporate co-precipitation sludges or evaporator concentrates ?

G. LEFILLATRE, France

At Valduc, we usually incorporate mixtures of 60 % in volume of chemical co-precipitation sludges (calcium phostates, aluminium and iron hydroxides) and diatomates, and 40 % in volume of evaporator concentrates at ~ 400 g/l of $NaNo_3$.

W. HILD, F.R. of Germany

What is the evaporation capacity for direct bituminization ?

G. LEFILLATRE, France

60 l/h of evaporated water with the LUWA L150 evaporator.

W. HILD, F.R. of Germany

What type of bitumen are you going to use in your Cadarache facility ?

G. LEFILLATRE, France

The Mexphalte 40/50.

W. HILD, F.R. of Germany

Have you noted degradation of NH_4NO_3 in your typical waste at Cadarache during incorporation into bitumen ? We have observed partial degradation, mainly in the final part of the extruder, of condensates containing nitric acid.

G. LEFILLATRE, France

No, we have incorporated concentrates with the thin film evaporator, with a NH_4NO_3 content of 39 g/l. We have not noted any significant change of the pH of distillates and we could not get any evidence concerning NH_4NO_3 decomposition.

Discussion

K. BRODERSEN, Danemark

Je crois comprendre que vous avez procédé à certains essais de lixiviation sur des blocs de bitume en vraie grandeur ; le taux de lixiviation obtenu au moyen de ces expériences est-il identique ou comparable aux taux de lixiviation mesurés à l'aide de l'essai type de l'AIEA qui s'effectue à l'échelle du laboratoire ?

G. LEFILLATRE, France

Oui, les résultats de lixiviation avec des enrobés grandeur nature sont égaux et même supérieurs à ceux obtenus avec des éprouvettes de laboratoire [cf. rapport CEA R3743 (1969)].

W. HILD, R.F. d'Allemagne

A Valduc, incorporez-vous des boues de co-précipitation ou des concentrats d'évaporation ?

G. LEFILLATRE, France

A Valduc, nous enrobons généralement des mélanges 60 % en volume de boues de co-précipitation chimique (phosphates de calcium, hydroxydes d'aluminium et de fer) et de diatomées et 40 % en volume de concentrats d'évaporation à ~ 400 g/l de $NaNO_3$.

W. HILD, R.F. d'Allemagne

Quelle est la capacité d'évaporation dans la bituminisation directe ?

G. LEFILLATRE, France

60 l/h d'eau évaporée avec l'évaporateur LUWA L150.

W. HILD, R.F. d'Allemagne

Quel genre de bitume allez-vous utiliser dans votre installation de Cadarache ?

G. LEFILLATRE, France

Le Mexphalte 40/50.

W. HILD, R.F. d'Allemagne

Avez-vous observé une dégradation de NH_4NO_3 dans vos déchets type à Cadarache pendant l'enrobage dans le bitume ? Nous avons observé une dégradation partielle, surtout dans la partie finale de l'extrudeuse, les condensats contiennent de l'acide nitrique.

G. LEFILLATRE, France

Non, nous avons enrobé des concentrats, avec l'évaporateur à couche mince, à des teneurs en NH_4NO_3 de 39 g/l. Nous n'avons pas observé de modification sensible du pH des distillats et nous n'avons pas pu mettre en évidence une décomposition du NH_4NO_3.

BITUMINIZATION OF RADIOACTIVE WASTES AT THE NUCLEAR RESEARCH CENTER KARLSRUHE - EXPERIENCE FROM PLANT OPERATION AND DEVELOPMENT WORK

W. Hild, W. Kluger, H. Krause
Gesellschaft für Kernforschung mbH
D-7500 Karlsruhe, Germany

Abstract

A summary is given of the main operational experience gained at the Nuclear Research Center Karlsruhe in 4 years operation of the bituminization plant for evaporator concentrates from low- and medium level wastes. At the same time some of the essential results are compiled that have been obtained in the R+D activities on bituminization.

Résumé

Un résumé est donné sur les principales expériences operationelles obtenues au Centre de Recherches Nucléaires de Karlsruhe en 4 années d'operation sur la bituminisation des concentrats d'évaporation des effluents de faible et moyenne activité. En meme temps on présente quelques résultats essentiels obtenus sur les activités de recherche et de developpement de l'enrobage des déchets radioactifs en bitume.

1. INTRODUCTION

Due to the very restrictive activity discharge limits, allowing release of liquid effluents to the main canal only after decontamination to activity concentration equal to drinking water standards for occupationally exposed personnel, efficient waste treatment techniques had to be selected for the Nuclear Research Center Karlsruhe. Evaporation of low and medium level waste streams turned out to be the most effective and most economic procedure from various liquid waste treatment processes tested on a pilot scale in the early sixties and has thus been adopted in a large treatment plant starting operation in 1968 {1}.

The evaporator concentrates from that plant were solidified during the first years by mixing with concrete. This batchwise operated solidification technique has successfully been replaced by the continuous bituminization process in 1972. The plant, the essential part of which is a self-cleaning screw-extruder evaporator, allows at the same time the effective evaporation of the water from the evaporator concentrates and the homogeneous incorporation of the radioactive salt residues into bitumen.

In comparison with the solidification by mixing with concrete incorporation into bitumen has - among others - the following main advantages

- the volume of the solidified waste that has to be stored perpetually is roughly 5 times smaller, and
- the leach rates of the bitumen products are generally two orders of magnitude lower.

The selection of the process type installed in the waste treatment facilities at Karlsruhe was based both on R+D-work that started in 1964 comprising lab-scale investigations and process engineering tests and on the evaluation of bituminization experience collected elsewhere {1}. Emphasis was given mainly to the definition and evaluation of process and product data relevant to the safe incorporation of the evaporator concentrates typical for the Research Center and to the safe intermediate storage, transport and final disposal in the salt mine Asse of the resulting bituminised waste products.

Operation of the industrial bituminisation plant is still flanked by R+D activities aiming both at the definition of operation conditions for the bituminization of other types of radwaste and at the investigation of related safety aspects. To this end and for trouble shooting experiments a bench scale screw extruder unit is operated too, that guarantees realistic incorporation experiments yielding bitumen products of exactly the same characteristics as those produced in the bituminization plant. This condition is essential for the practical application of the experimental results obtained.

This paper is an attempt to summarize some of the essential experience and results gained in radwaste bituminization at Karlsruhe both in plant operation and R+D activities.

2. PLANT OPERATION

2.1 Description of the bituminization plant

A detailed description of the bituminization plant, that has been installed in existing rooms of the waste treatment facilities, is given in reference {2}. For convenience figure 1 shows a simplified functional flowsheet. Bituminization occurs in a Werner & Pfleiderer type ZDS-T 120 extruder evaporator, with a pair

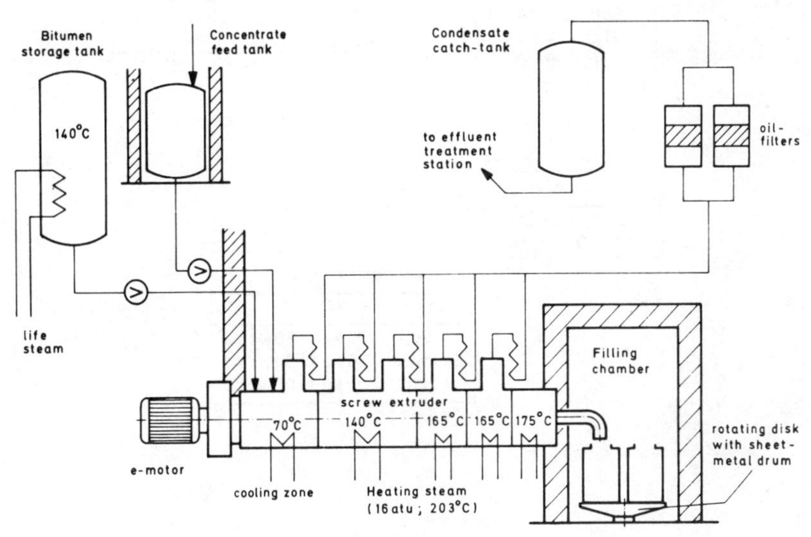

Simplified functional Flowsheet

Figure 1

of intermeshing screws of self-cleaning profile. Liquid standard grade Mexphalt 15 bitumen that is stored at 140°C in a 20 m^3 tank is fed to the extruder together with evaporator concentrate from the shielded 1 m^3 concentrate feed tank. At the typical temperature profile stated evaporation of water and incorporation of the radioactive residues takes place simultaneously. After condensation the water passes an oilfilter and is recycled to the effluent evaporator. The bitumen product flows into one of six 175 l sheet-metal drums standing on a turntable in the filling chamber. Extruder and bitumen tank are heated by steam.

It should be noted that this plant was the first plant utilising this particular technology for the direct bituminization of evaporator concentrates with up till 70 wt.% of the contained salts being $NaNO_3$.

2.2 Plant performance

Figure 2 shows the flowsheet representing normal plant operation, with the screw extruder working at optimum evaporation capacity of roughly 140 kg H_2O/h yielding an extremely homogeneous bitumen product of 50 wt.% salt with $\leq$ 0.5 wt.% residual water. The PH of the evaporator concentrate has to be adjusted between 8 and 10.

Incorporation Flowsheet

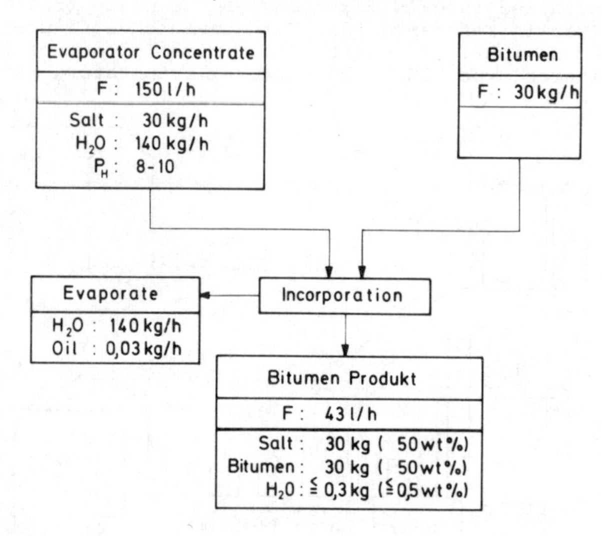

Figure 2

Since start-up of the plant in 1972 roughly 700 m^3 of evaporator concentrates from more than 40,000 m^3 low and intermediate level effluents produced in the Center and the 40 t/y reprocessing plant WAK were solidified by incorporating the solid residues containing some 10^4 Curies into bitumen. Approximately 2000 drums were produced with an average specific activity of 100 Ci/m^3 and 50 wt.% salt content and were shipped to the Asse salt mine for disposal. Distribution of dose rates was as follows: roughly 50%, 30% 10% and 5% of the drums had surface dose rates of up to 20, 40, 60 and 80 R/h, respectively, whereas the rest showed dose rates up till 200 R/h. To cope with this the sheet-metal drums were either placed into a 200 l reinforced drum that was inserted in special reusable shielded transport containers or the sheet-metal drums were inserted into prefabricated concrete containers which after being topped of with concrete were transported to the Asse for disposal.

From the operation experience that is described in detail in references {2,3} let us cite only the following points:

Bitumen losses and coking problems known from bitumen storage at ≥ 200°C in refineries were not observed in the bitumen storage tank (140°C). The strainer in front of the bitumen metering pump becomes mainly clogged by coking sediments one day after recharging the bitumen tank.

Sedimentations of salt and bitumen salt mixes occasionally occurred at the steam domes but could successfully be cleaned by a specially developed cleaning system operating with steam jets.

To cope with the volume contraction upon cooling and to provide for better cooling during filling the 6 drums on the turntable are filled successively for approximately 30 minutes each, and are then changed. Thus a drum is filled after 9 to 10 changes. After an additionary cooling period of about 24 hours when the centerline temperature of the products has dropped to $\leq$ 110°C the drums are transferred to the interim storage area. The outside of the sheet metal drums contains always some quite adhering contamination, which is not eliminated as the drums are anyhow inserted into a second container (reinforced steel, or concrete drum). Due to this second containment and due to the fact that shortly after filling radioactive aerosols are no longer emitted from the product capping of the metal sheet drum has been abandoned.

Air contamination level in the filling chamber was found to be 10 times MPC for α and 50 times MPC for β when processing an evaporator concentrate of 90 Ci/m^3.

An overall decontamination factor of roughly 6000 was found for the distillate from the extruder when processing the above mentioned 90 Ci/m^3 concentrate. This DF holds also for the individual nuclides as Cs-137, Ru-106, Ce-144 and Sb-125, demonstrating a rather good decontamination during normal operation. No further measurable decontamination of the condensates takes place in the oil filter. During plant operation the activity increases, however, in the filter medium by cumulation.

2.3 Off-design conditions

Occasionally the bitumen product discharge became clogged due to material build-up in cavities of the discharge compartment. If the extruder is not shut down in time back pressure forces the product up into the last steam dome, where the continuously produced steam can foam it and force it up as far as the condensers. In such cases the steam dome and discharge compartment have to be dismantled for cleaning. Although these build-ups seem to be related to off-standard feed ratios entire clarification has not been obtained. Clogging can however be avoided by inspection of the product outflow and it is completely excluded by a constructional change of the discharge part into forced discharge utilising the transporting action of the extruder.

After roughly 7.500 hours of operation the screws were withdrawn and replaced. An abrasion of up to 7 mm was observed at some of the transporting - and mixing - and kneading disks. Obviously this phenomenon is mainly due to erosive wear which is the more pronounced the higher the salt content of the product. Especially at off-standard feed ratios leading to bitumen depleted products the lubrificating action of the bitumen is drastically decreased. Corrosion can be excluded under normal operation conditions; when, however, processing evaporator concentrates that exceptionally have a too low or too high PH corrosion might play a role, too. Actually tests are performed by the manufacturer Werner & Pfleiderer, with different screws aiming at the determination of a construction material with higher resistance to erosion.

In the course of 4 years of operation, two incidents occurred in the filling chamber. In both cases the bitumen product was foaming and fumes were detected that suddenly caught fire. As a result of this the drums that had received fractions of the corresponding bitumen product caught fire, too (1 drum in the first, 2 drums in the second case). The other drums (5 and 4 respectively) did not burn. Both fires were under control within a short time and could be extinguished with CO_2. No radioactivity escaped from the plant, no person was injured and only little damage and contamination occurred in the filling chamber. In the first incident the fire was caused by ignition of the vapors of organic solvents contained in the waste concentrate. In the other incident, an evaporator concentrate was processed having a pH of 13.8 instead of the standard value 8 to 10. At the same time, the agitator of the concentrate feed tank had failed and a separation of organic compounds contained in the waste (TBP, degradation products, antifoaming agents polyethyleneoxide adducts) could take place. As demonstrated in experiments with the bench scale unit these products are decomposed into easily flammable volatile compounds during incorporation at high alkalinity. The resulting bitumen products had ignition points around 200°C, whereas at pH 8 to 10 no degradation occurs and ignition points of the product are in the region of 400°C. After appropriate pH adjustment with H_3PO_4 the remaining evaporator concentrate was incorporated without any difficulty.

Finally a reaction took place in one bitumen drum in the interim storage bunker. Investigations showed that this drum received a slightly foaming bitumen product that was processed from the final fraction of a non agitated evaporator concentrate in the feed tank containing large amounts of PVC-powder, that - as turned out later - is a filler in pickling pastes used for decontamination and that floats on the surface of evaporator concentrate. This drum-being the last prior to weekend shut down of the plant - was in contrast to the operation instructions not allowed to cool down during 24 hours but immediately after filling with the 200°C bitumen product inserted into a 200 l reinforced drum. This assembly acted as a thermostate and the PVC continued to decompose yielding HCl that most probably initiated an oxidation of the organic material by decomposing nitrates. Although the escaping fumes set an alarm no fire occurred. Apart from the sealing no damage was detected at the reinforced drum and the incident had no influence on the other drums stored in the immediate vicinity. It's worthwhile mentioning that due to this event a new pickling paste with an inorganic filler was developed that is now exclusively being used at Karlsruhe.

2.4 Process optimisation

Due to the fact that the incidents mentioned in section 2.3 were all in relation with inhomogeneities in the evaporator concentrates measures have been taken for plant shut down whenever the agitator of the concentrate feed tank fails. PH control and adjustment to the mentioned range 8 to 10 is mandatory for operating the bituminization plant. In addition to this DTA analysis of the dry residue of the evaporator concentrate and a 1:1 mixture with bitumen is an excellent indication for eventual off standard samples asking for special treatment.

The electrical installation of the filling chamber is now explosion-proof. The venting system guarantees a fiftyfold air renewal in the filling chamber per hour. Temperature gauges are installed that register any abnormal rise in temperature. Furthermore fume and gas detectors are installed (CO and CH_4 calibrated) that indicate any flammable gas composition in the filling chamber atmosphere. The latter are linked with an automatic fire-extinguishing equipment containing 1.5 t CO_2.

The outlet part of the screw extruder has been modified to avoid clogging of the discharge compartment. In addition inspection of the product discharge is maintained.

Modification of the oil filter for use of filter cartridges that are rejected after saturation instead of reusing the filter housing, leads to an increase in service life of the filter media. The use of filter cartridges substantially reduces the time involved for the filter change, thus allowing higher specific activities on the filters.

Measures are taken for efficient separation of organic solvents from the low and intermediate level effluent streams prior to evaporation (in the vapor compression evaporator).

Operation temperature in the screw extruder is usually limited to 180°C and does never exceed 200°C.

Filled drums are allowed to cool for further 24 hours bringing the centerline temperature of the drums down to the softening point of the product ($\sim$100°C). After this cooling period the drums are transported to the interim storage bunker.

Prior to start up and after shut-down of the bituminization of evaporator concentrates the extruder is operated only with bitumen for a period of roughly 15 minutes.

With these measures the bitumen plant has continued extremely reliable operation.

3. RESEARCH AND DEVELOPMENT

As already indicated both the conception of the plant and the definition of the operation conditions were based on the results of extensive R + D work. The bitumen selected for the plant is a standard grade Mexphalt 15 or Ebano 15 medium-hard distilled bitumen of 67°C to 72°C ring and ball softening point and a penetration of 10 to 20. Within the existing standard grades this type of bitumen has the highest flash point (> 290°C), another important reason for its selection.

Due to the fact that approximately 30% of the low-level effluents ($\sim 10^{-2}$ Ci/m^3) and more than 80% of the intermediate level effluents ($\leq$ 100 Ci/m^3) result from the reprocessing plant, they contain rather high $NaNO_3$ concentrations. In fact the $NaNO_3$ content in the salt residues of the evaporator concentrates actually processed amounts to up till 70wt%. This is why the largest part of the investigations were strictly devoted to the safety aspects related to both the processing of such high nitrate containing evaporator concentrates and to the resulting products. The problem of bitumen oxidation and hardening was thus thoroughly investigated {4}. It could be concluded that hardening and oxidation of the bitumen can safely be avoided by incorporating at slightly alkaline conditions, at temperatures $\leq$ 200°C and preferrably at short residenc times. The latter two conditions are best met in a screw extruder.

3.1 Thermal stability

One of the safety aspects that has received considerable attention was the thermal stability of the products. An exhaustive investigation was performed at the Institute for the Chemistry of Propellants and Explosives (ICT) on the thermal and mechanical sensitivity of bitumen/oxygen salt mixtures {5}. Investigations were performed mainly with bitumen products of 60 wt.% nitrates from various metals and other additives that were expected to be potential constituents of evaporator concentrates. The high salt content compared with that of the actual products prepared in the plant ($\sim$50 wt%) has purposely been selected to be on the safe side. Bitumen products and pure bitumen showed the same behaviour on 40 hours heating at 100°C under vacuum, demonstrating good compatibility of the salt mixtures with bitumen. Ignition temperatures of the products were above 400°C, only some samples with increased fractions of calcium-nitrate and transition metal nitrates showed ignition temperatures between 360° and 380°C. Combustion behaviour studied in the red hot steel pan (~700°C) showed a faster time to ignition for the bitumen products (1 to 5 sec) than for the pure bitumen (30 sec); the burning periods were of the same order of magnitude, with the bitumen products burning more briskly than the pure bitumen. Rapid heating under contained conditions demonstrated that there was no hazard of explosion. Furthermore the products were not sensitive to mechanical stresses and detonation stresses. The investigations thus demonstrated, that the products investigated do not belong to the category of substances involving an explosion hazard.

In addition 175 l sheet-metal drums containing 57:43 wt% bitumen: $NaNO_3$ products were ignited by powerful oilfires in field tests to study the combustion behaviour and fire fighting measures {6,7}. The tests confirmed, that the products cannot be considered easily flammable. During the first 10 minutes the combustion was peaceful, afterwards partially vigorous (obviously due to the decomposition of $NaNO_3$) but never leading to a deflagration or explosion. Total combustion time for 211 kg bitumen product (87% filling of the 175 l drum) was 85 minutes, roughly 27% of the $NaNO_3$ was carried with the fumes most probably as Na_2O aerosols. Fire fighting tests have demonstrated that bitumen product fires can be brought under control without difficulties and that CO_2 is the most suitable fire extinguishing medium (see sections 2.3 and 2.4). Combustion tests will be completed by exposing bitumen product packages (175 l sheet metal drum in a 200 l reinforced drum or a prepfabricated concrete container, see section 2.2) to oil fires.

The question whether bitumen products might catch fire as a consequence of an eventual ignition of an hydrogen - air mixture resulting from radiolysis (see section 3.3) is actually under investigation at the ICT. First tests in a simulated small scale storage chamber demonstrated that ignition of various H_2/air mixtures with H_2 contents between 4.3 and 10.5 vol.% did not lead to an inflammation of the bitumen product test-specimen. Although these tests demonstrate again that the bitumen products are not easily flammable, full scale gas explosion tests with a couple of 175 l sheet-metal drums containing bitumen products will be performed this year in a special blasting bunker.

Differential thermal analysis (DTA) is an excellent tool for the determination of the thermal stability of chemical compounds. DTA curves of active bitumen product samples from the plant show all endothermic peaks in the temperature range between 270°C and 300°C. Exothermic peaks are generally observed in the region of 400°C. It could be shown that the endothermic peaks indicate the melting point of $NaNO_3$ (306°C) which due to other salt impurities can considerably be lowered {8}. The exothermic peaks around 400°C coincide generally with the burning points of the mixtures, as determined by a non standardized test procedure where 1 g of product is heated on a steel plate in presence of a fanning flame. Endothermic peaks in the region of the operation temperatures of the extruder generally indicate volatile organic compounds. The antifoaming agents, for instance, that were originally utilised in the vapor compression evaporator showed marked endothermic peaks at 130°C and have been replaced by a more stable compound. Due to the significant information on the thermal behaviour DTA tests are routinely performed both in plant operation and R+D-activities.

The above mentioned test procedure, although not standardized, has proven to be a very reliable method since starting investigations on thermal stability. Testing various bitumen grades, lowest burning points (260° to 300°C) were found for Mexphalt R85/40 or R 90/40 blown bitumen and highest burning points (410°C to 440°) were found for standard grade bitumen Mexphalt 15 or Ebano 15. Various nitrate salt mixtures with the latter bitumen grades showed burning points well above 300°C, generally in the region of $\geq$ 400°C {9}, being in excellent consistency with the values measured by ICT {5}. Burning points of 50 wt.% bitumen products obtained in the bench scale unit from simulated power reactor wastes (spent mixed bed ion exchanger, evaporator concentrates with high boric acid - or high tenside content) showed values well above 400°C {10}.

In connection with the reconnaissance of the incident mentioned in section 2.3 a simulated evaporator concentrate solution containing 175 g/l $NaNO_3$, 5 g/l $NaNO_2$ and 20 g/l of a 1:1 mixture of dibutyle- and monobutyle phosphoric acid was incorporated at pH 11 and pH 13.5 in the bench scale unit yielding 50 wt.% bitumen products. Operation temperatures were as in the production plant. In contrast to the product of low alkalinity, the product of high alkalinity was vigorously foaming. The burning point of the product was at 200°C whereas the product of low alkalinity had a burning point of 400°C. In contrast to this result the burning point of the product was not lowered when incorporating an evaporator concentrate with only 1.5 g/l DBP/MBP at high pH. This and similar experiments demonstrated that at high alkalinity organic compounds can be decomposed into easily flammable compounds at the operation temperatures of the extruder. Care has thus to be taken for a homogeneous feed and a pH adjustment between 8 and 10.

3.2 Resistance to leaching

Apart from thermal stability resistance to leaching is another important property needed for the characterization of solidified radioactive wastes. Leach tests were performed with numerous bitumen products containing roughly 40 wt.% salts of representative composition for the evaporator concentrates of the Research Center {4,11}. Leaching was measured in distilled water for one year by determining the amount of sodium leached. The average leach rate of 16 different products was $5 \cdot 10^{-4} \cdot g \cdot cm^{-2} \cdot d^{-1}$ with $1 \cdot 10^{-5}\ g \cdot cm^{-2} \cdot d^{-1}$ being the lowest and $2 \cdot 10^{-3} g \cdot cm^{-2} \cdot d^{-1}$ the highest value. The simultaneous determination of sodium, iron and calcium did not reveal any significant differences in the specific leach rates of these ions.

A direct proportionality of the leach rate with the amount of salt incorporated was found for bitumen-$NaNO_3$ products as indicated in table I.

Table I Leach rates of bitumen / $NaNO_3$ products as a function of $NaNO_3$ concentration (average values over 1 year)

$NaNO_3$ content {wt.%}	0.1	1	5	10	20	38.5
leach rate { $g \cdot cm^{-2} \cdot d^{-1}$ }	not detectable	not detectable	$4 \cdot 10^{-6}$	$7 \cdot 10^{-6}$	$2 \cdot 10^{-5}$	$9 \cdot 10^{-5}$

Another interesting result is the relation between the leach rate and the dispersion of the salts in the bitumen products. In experiments with bitumen products containing 38.5 wt.% NaCl of defined particle size it was demonstrated that products with coarser crystals are more easily leached than products with finer crystals. After one year the product samples with crystals of the sieve fraction 0.5 - 0.8 mm showed a cumulative leaching between 0.7 and 1% compared with the 0.1% value of the product samples with crystals of the sieve fraction 0.05 - 0.08 mm. As the screw extruder leads to extremely homogeneous products with average particle sizes between 10 and 30µ these bitumen products are by one order of magnitude less leachable than comparable products from pot processes.

Products of bad leach resistance have been found when incorporating sodium carbonate into bitumen. A product containing for instance, roughly 19 wt.% $NaNO_3$ and 19 wt.% Na_2CO_3 had an average leach rate (1 year) of $2 \cdot 10^{-3}$ g $\cdot$ $cm^{-2} \cdot d^{-1}$. For a bitumen product containing 38,5 wt.% Na_2CO_3 an average leach rate of $1 \cdot 10^{-2}$ g $\cdot$ $cm^{-2} \cdot d^{-1}$ was measured during 84 days. At the same time these products showed considerable swelling when stored under water. The 38.5 wt.% Na_2CO_3 sample had after 14 days immersion approximately doubled its original volume. This anomalous behaviour was also noticed in the shape of the leaching curve that, in contrast to normal findings showed after 3 days immersion in water a sharp linear increase of the differential leach rate during a period of one week.

Similar exceptional leaching behaviour was found for bitumen products containing Na_2SO_4. As for the Na_2CO_3 samples an increase in the differential leach rate was observed. Cumulative leaching during 7 days from ~ 10 cm Ø and ~ 7 cm high samples amounted to 49.2%, 10.9% and 1.2% for bitumen products containing 50 wt.% 39 wt.% and 30 wt.% Na_2SO_4. The products suffered from swelling, too, as indicated by pointed cone-shaped elevations forming after 7 days immersion in water. As in the case of the Na_2CO_3 samples, this behaviour is most probably due to the fact that both Na_2CO_3 and Na_2SO_4 in their anhydrous form tend to bind considerable amounts of water of crystallization. This reaction leads in turn to a pronounced increase in the volume of the crystals, that provokes crevices in the bitumen product thus enlarging the entrance of further water into the bitumen matrix.

During an experimental program performed with the bench scale unit to define operation conditions for the bituminization of power reactor wastes, some products were found that did not completely correspond with the good leaching characteristics of bitumen products normally encountered {10}. A bitumen product containing 53 wt.% bitumen and 47 wt.% salt residues of an evaporator concentrate composed among other constituents mainly of boric acid (150 g/l H_3BO_3) alkalized by 82 g/l NaOH to pH 12, showed an average leach rate over 100 days of $7 \cdot 10^{-3}$ g $\cdot$ $cm^{-2} \cdot d^{-1}$. Another bitumen product containing 50 wt.% salt residue, more than 2/3 of which were detergents, soap powder and laundry cleansing agents revealed an average leach rate of $1.5 \cdot 10^{-2}$ g $\cdot$ $cm^{-2} \cdot d^{-1}$ during 100 days leaching. The products showed some swelling upon immersion in water, too. This behaviour is due to the high alkalinity and the large amounts of detergents that have an emulsifying and dispersing action.

An amelioration of the resistance to leaching can be obtained either by chemical pretreatment of the concentrates prior to incorporation or by coating the final products with a layer of pure bitumen. A 38.5 wt.% $NaNO_3$ containing bitumen product covered with a 5 mm thick coating of pure bitumen had for instance an average leach rate of $3 \cdot 10^{-8} \cdot$ g $\cdot$ $cm^{-2} \cdot d^{-1}$ after 5 years continuous leaching i.e. a leach rate that is almost 4 orders of magnitude lower than that of the uncoated product. A 40 wt.% Na_2SO_4 containing bitumen product with a similar coating did not swell at all when immersed in water. The same was true for a bitumen product containing 17 wt.% Na_2SO_4, 17 wt.% $CaSO_4$ and 14 wt.% NaCl, i.e. a product that has been obtained by incorporating a Na_2SO_4 solution where half of the Na_2SO_4 had previously been precipitated by $CaCl_2$. Leach tests are still continuing, indicating drastically reduced leach rates. These examples demonstrate, that products of adequate leaching behaviour can successfully be obtained by appropriate chemical pretreatment and/or coating.

3.3 Stability towards radiation

For the performance of external radiation tests two radiation sources have been utilized: A linear accelerator for 10 MeV electron pulses (frequency 170 sec^{-1}, time 5 μsec) with an "apparent" dose rate of roughly $8 \cdot 10^6$ rad/h and an irradiation facility in a fuel element storage pond with an average γ dose rate of 10^5 R/h.

First irradiation tests with various bitumen types and different bitumen/salt mixtures (61.5/38.5 wt.%) indicated that a marked increase in the softening point and thus a change in the viscoelastic properties of bitumen and bitumen/salt mixtures can first be noticed at irradiation to $\geq$ 100 Mrad {4}

Irradiation tests with distilled bitumen (Mexphalt 15) blown bitumen (Mexphalt R 90/40) and their corresponding products containing 50 wt.% $NaNO_3$, were performed with 10 MeV electrons from the linear accelerator {12}. When irradiated to $5 \cdot 10^8$rad the blown bitumen and its corresponding product showed a very pronounced increase in the softening point (90°C to 117°C and 113°C to 159°C, respectively). When heating the blown bitumen during 5 hours to 150°C prior to irradiation this increase was still more pronounced (117°C to 155°C). For the distilled bitumen and its corresponding product a rather small increase was found (77°C to 83°C and 101°C to 108°C). All pure bitumen samples and their corresponding products showed marked swelling after irradiation to $5 \cdot 10^8$rad: Porosities ($\varepsilon = 1 - \frac{\rho'}{\rho}$) (1) after irradiation were between 0.1 (Mex 15) and 0.4 (Mex. R90/40) and around 0.3 for the two corresponding products. Hydrogen production at an irradiation to $1 \cdot 10^8$ rad was 0.71 cm^3/g (Mex. R90/40) and 0.56 cm^3/g (Mex 15) for the pure bitumen samples whereas roughly half these values were found for the corresponding products 0.33 cm^3/g and 0.24 cm^3/g. It is interesting to note that a 5 hours heat treatment at 150°C prior to irradiation lowers the hydrogen production of the bitumen by about 10%.

A bitumen product containing 57.2 wt.% Ebano 15, 42 wt.% $NaNO_3$, 0.75 wt.% H_2O and 0.05 wt.% NaOH, that was produced during the cold start up of the bitumen plant showed porosities of 0.006, 0.14, 0.22 and 0.26 after irradiation with 10 MeV electrons to total absorbed doses of 10^7, $5 \cdot 10^7$, 10^8 and $5 \cdot 10^8$ rad respectively {6}. For both irradiation with 10 MeV electrons and γ radiation from the fuel element storage pond a linear relation was found between the amount of H_2 formed and the absorbed dose within the 100 Mrad dose range investigated. The specific hydrogen production rate determined was between 0.4 and 0.5 cm^3/g and 100 Mrad {7}. Though in the same order of magnitude as the values cited above this value is somewhat higher. As the hydrogen production has a certain influence on the storage conditions (H_2 accumulation in the atmosphere) further tests will be performed by studying the radiation stability with internal radiation from incorporated radionuclides in cooperation with EUROCHEMIC.

---

(1) ρ': density after irradiation, ρ: density without irradiation

Irradiation of the bitumen products produced from spent power reactor ion exchange resins and simulated evaporator concentrates rich in borates and detergents (see sections 3.1 and 3.2) {10} showed porosities between 0.01 and 0.04 for irradiation with both 10 MeV electrons and γ radiation to a total integrated dose of 80 Mrad. Hydrogen production was 0.3 cm^3/g for 80 Mrad.

As already mentioned the investigations of the radiation stability will be terminated in cooperation with EUROCHEMIC by internal irradiation tests utilising representative bitumen products with incorporated fission products and α-emitting transuranium elements.

4. CONCLUSION

In concluding it can be stated that both plant and R+D experience demonstrate that bituminization is a safe and effective means for the solidification of low and intermediate level radioactive wastes. The screw-extruder technique proved to be very reliable and flexible for continuous incorporation to extremely homogeneous products.

Escorting R + D work by operating a bench-scale unit of exactly the same features as the plant and by characterizing the products obtained for their thermal stability, leachability and radiation stability has proven to be a very valuable supplement both with respect to plant operation and to the definition of incorporation conditions for new types of radwaste.

From a radiation resistance point of view bituminization has, however, to be limited to integrated total doses of 10^8 to 10^9 rads, a dose range corresponding to the category of medium level wastes. In this connection, and due to the experience gained, there is no doubt that bituminzation is very well suited for the solidification of waste concentrates from power reactors and will thus find a broad application in the future. As far as medium level waste streams from reprocessing plants are concerned work is continuing to still further increase the safety of this process by applying processes that reduce both the activity and the nitrate content from these wastes {13}.

5. REFERENCES

{1} Bähr, W. and al.: "Experiences in the treatment of low- and intermediate-level radioactive Wastes in the nuclear research center, Karlsruhe", Proc. Symp. Management of Low- and Intermediate-Level Radioactive Wastes, p. 461 IAEA, 1970

{2} Meier, G. Bähr, W.: "Die Fixierung radioaktiver Abfälle in Bitumen. Teil 1 Die Betriebsanlage zur Fixierung radioaktiver Verdampferkonzentrate in Bitumen im Kernforschungszentrum Karlsruhe" German report KFK 2104, (april 1975)

{3} Bähr, W., Hild, W. Kluger, W.: "Bituminization of Radioactive Wastes at the Nuclear Research Center Karlsruhe", German report KFK 2119 (October 1974)

{4} Kluger W. and al.: "Fixing of Radioactive Residues in Bitumen", German Report KFK 1037, (august 1969)

{5} Backof, E. Diepold, W.: "Study of the Thermal and Mechanical Sensitivity of Bitumen/Oxygen Salt Mixtures" German Report KFK-tr-450 (July 1975)

{ 6} Krause, H. editor:"Jahresbericht 1972" German report KFK 2000, p. 10 (november 1974)

{ 7} Krause, H. Rudolph, G. editors: "Jahresbericht 1973" German report KFK 2126, P. 14, (May 1975)

{ 8} van Artsdalen, E.R.: "Complex ions in molten salts. Ionic association and common ion effect", Journ. Phys. Chem. vol LX, p. 172, (1956)

{ 9} Krause H. editor "Jahresbericht 1970" German report KFK 1500, p. 9 (June 1972)

{10} Hild, W. and al.: "Verfestigung radioaktiver Abfallkonzentrate aus Leistungsreaktoren durch homogene Einbettung in Bitumen", Kolloqiumsreferat D3/04 NUCLEX Basel (October 1975)

{11} Krause, H., editor: "Jahresbericht 1969", German report KFK 1346, p. 16 (January 1971)

{12} Krause, H. editor: "Jahresbericht 1971" German report KFK 1830, p. 7 (June 1973)

{13} Bähr, W. Hild, W. and al: "Recent Experiments on the Treatment of Medium Level Wastes and Spent Solvent and on Fixation into Bitumen" IAEA/NEA Intern. Symposium Managm. Rad. Wastes from Nucl. Fuel Cycle, Paper No. IAEA-SM-207/81 (March 1976)

Discussion

J.E. STEWART, United States

If nuclear power plants used calcium hydroxide rather than sodium hydroxide (as Eurochemic does at Mol) for neutralization, would leachability rates be improved ?

W. HILD, F.R. of Germany

You refer to waste concentrates containing Na_2SO_4 which showed bad resistance to leaching. If you precipitate the SO_4^{--} by Ca^{++} ions, you form $CaSO_4$ and this compound leads to products of excellent leach resistance when incorporated in bitumen. It is in fact not neutralisation we were talking about, it is chemical precipitation. You are, however, right if you think of neutralizing the H_2SO_4 - containing solutions resulting from the regeneration of ion exchangers at nuclear power stations. There neutralisation by $Ca(OH)_2$ would immediately lead to the compound desired.

N. FERNANDEZ, France

In the Federal Republic of Germany, and particularly at Karlsruhe, you incorporate low level radioactive concentrates in bitumen and store them later in a solid formation. Do the German authorities consider it absolutely necessary to combine, for the management of low level radioactive material which are practically free of alpha emitters :

1) the high degree of containment provided by incorporation into bitumen, and

2) the high degree of containment of particular geologic formations (salt for example) ?

W. HILD, F.R. of Germany

The philosophy in the Federal Republic of Germany has always been to protect as much as possible the environment from radioactive contamination. This is why at Karlsruhe the effluents are evaporated and the evaporated concentrates solidified with concrete and, since 1972, with bitumen. This last solution was chosen first to reduce the volume of waste to be transported to the Asse Salt Mine, secondly to arrive at products with more favorable leaching characteristics, and thirdly to use a continuous process.

N. FERNANDEZ, France

You have mentioned the causes of the two fires of bitumen products at Karlsruhe (presence of solvents, organic products, TBP ; very high pH, etc.). Do you not think that it is essential to always foresee in this field abnormal operation and to consequently provide the incorporation facilities with electric explosion-proof equipment, with important air ventilation to avoid possible accumulation of combustible vapours, as well as means to stop immediately an abnormal reaction (by adding water, CO_2, etc.).

W. HILD, F.R. of Germany

You are absolutely right. We consider the incidents mentioned as unpleasant but nevertheless very useful experiments, and as you can read in my paper, we have since then taken all the technical precautions you have indicated. Since these measures have been adopted, the plant continues operations without problems.

Discussion

J.E. STEWART, Etats-Unis

Si les centrales nucléaires utilisaient de l'hydroxyde de calcium au lieu d'hydroxyde de sodium (comme la Société Eurochemic à Mol) pour la neutralisation, les taux de lixiviation s'en trouveraient-ils améliorés ?

W. HILD, R.F. d'Allemagne

Vous vous référez à des concentrats de déchets contenant du Na_2SO_4 qui ont accusé une mauvaise résistance à la lixiviation. Si vous précipitez le SO_4^{--} par des ions Ca^{++}, vous obtenez la formation de $CaSO_4$ et ce composé conduit à des produits présentant une excellente résistance à la lixiviation lorsqu'ils sont incorporés dans du bitume. En fait, ce n'est pas de neutralisation dont nous parlons, mais de précipitation chimique. Cependant, vous avez raison si vous pensez à la neutralisation de solutions contenant du H_2SO_4 qui proviennent de la régénération des échangeurs d'ions dans les centrales nucléaires. Dans ce cas, la neutralisation par le $Ca(OH)_2$ permettrait d'obtenir immédiatement le composé souhaité.

N. FERNANDEZ, France

En Allemagne fédérale et à Karlsruhe en particulier, vous confinez les concentrats de faible activité dans le bitume pour les stocker ensuite dans les couches géologiques salines. Les autorités d'Allemagne fédérale jugent-elles nécessaire et indispensable d'associer pour la gestion des déchets radioactifs de faible activité et pratiquement exempts d'émetteurs "alpha" :

1) le haut degré de confinement de l'enrobage au bitume, et

2) le haut degré de confinement de couches géologiques particulières (salines entre autres) ?

W. HILD, R.F. d'Allemagne

La situation et la philosophie en Allemagne fédérale ont toujours été de protéger autant que possible l'environnement contre la contamination par des radioéléments. C'est ainsi que l'on a utilisé à Karlsruhe l'évaporation des effluents et la solidification des concentrats d'évaporation dans le ciment et, depuis 1972, dans le bitume. Cette dernière solution a d'ailleurs été choisie, premièrement pour diminuer le volume de déchets à transporter à la mine de sel de Asse, deuxièmement pour arriver à des produits avec des caractéristiques de lixiviation plus favorables et troisièmement pour arriver à un procédé continu.

N. FERNANDEZ, France

Vous avez indiqué les causes des deux incendies d'enrobé à Karlsruhe (présence de solvant, produits organiques, TBP ; pH très élevé, etc.). Ne pensez-vous pas qu'il est indispensable de toujours prévoir dans ce domaine un fonctionnement anormal et qu'en conséquence les installations d'enrobage doivent avoir un équipement électrique antidéflagrant, un renouvellement d'air important pour éviter les accumulations éventuelles de vapeurs combustibles, ainsi que des moyens pour arrêter instantanément une réaction anormale (par addition d'eau, de CO_2, etc.).

W. HILD, R.F. d'Allemagne

Vous avez tout à fait raison. Nous considérons les incidents mentionnés comme des expériences pénibles mais néanmoins très utiles et, comme vous pouvez le constater à la lecture de ma communication, on a, entre temps, pris toutes les précautions techniques que vous avez citées. Depuis ces mesures d'amélioration, l'usine a continué son travail sans problème.

RADIOLYSIS AND TEMPERATURE EFFECTS IN CASE OF UNDERGROUND STORAGE OF BITUMEN

E. Smailos, W. Diefenbacher, E. Korthaus, W. Comper
Gesellschaft für Kernforschung mbH, Karlsruhe
Federal Republic of Germany

Abstract

On conservative assumptions limit values were determined for the specific activity in the bitumen products, which safely avoid under the conditions of storage in a non-ventilated prototype cavity both the formation of an ignitable radiolytic gas/air mixture and intolerable heating of wastes products over the entire storage time

Depending on the filling factor of the cavity and on the age of fixed fission products, the limit values of the specific activity in the waste products allowing to avoid an ignitable gas/air mixture in the cavity range from 0.09 Ci/l to 0.78 Ci/l. The respective limit values of specific activity allowing to avoid intolerable heating of wastes (> 70°C) range from 0.3 Ci/l to 0.7 Ci/l, depending on the age of fission products and on the type of storage.

Résumé

Sur la base d'hypothèses conservatrices on a fixé des limites de l'activité spécifique dans les produits bitumeux éliminant avec certitude la formation d'un mélange gaz/air inflammable aussi bien que le chauffage inadmissible de déchets pendant toute la période de stockage sous les conditions de stockage dans une caverne prototype non-ventilée.

Selon le facteur de remplissage de la caverne et l'âge des produits de fission liés les limites de l'activité spécifique dans les déchets permettant d'eviter la formation d'un mélange gaz/air inflammable dans la caverne varient entre 0,09 Ci/l et 0,78 Ci/l. Les limites correspondantes de l'activité spécifique permettant d'éviter le chauffage inadmissible des déchets(> 70°C) vont de 0,3 Ci/l à 0,7 Ci/l suivant l'âge des produits de fission et le mode de stockage.

1. THE PROBLEM AND THE OBJECTIVES OF THE INVESTIGATIONS

In cooperation between Gesellschaft für Strahlen- und Umweltforschung mbH Munich (GSF) and Gesellschaft für Kernforschung mbH Karlsruhe (GfK) a prototype cavity of 10,000 m^3 volume is being constructed in the Asse II salt mine for trial storage of medium level wastes. Fig. 1 shows a drawing of the cavity. Under the technique of filling envisaged here the fixed waste products are lowered about 918 m from the surface down to the unloading facility by means of a shaft hoisting equipment and introduced into the cavity from this level in a free fall of about 70 m at the maximum.

Prior the disposal of bituminized radioactive wastes in the prototype cavity GfK had performed investigations into nuclear safety. It was studied in this context, whether radiolysis and heat problems respectively, arise in the prototype cavity as a result of radiolytic decomposition of bitumen and decay heat of the fixed fission products, which set a limit to the specific activity of the bitumen products when stored in the cavity.

The aim is pursued in the investigations of defining the limit values for the specific activity in bituminized wastes in which under the conditions of storage in a non-ventilated prototype cavity both the formation of an ignitable radiolytic gas/air mixture and intolerable heating of waste products during the entire period of storage are safely avoided. As a maximum admissible temperature for the bitumen products was assumed the softing point of bitumen Mexphalt 15 (70 °C).

2. ON THE FORMATION OF RADIOLYTIC GASES FROM BITUMINIZED WASTES

2.1 Model Assumptions and Computational Methods

The following assumptions were made for the calculations:

Type of Waste: Fixed LWR-fission products (33,000 MWd/t, 30 MW/t); bitumen Mexphalt 15/$NaNO_3$ products (50 wt. °/o salts), density 1.5 kg/l; specific activity of products at the time of filling 0.1 Ci/l up to 1 Ci/l; age of fission products (at the time of filling) 0.5 year, 1 year, 2 years.
Cavity and Type of Filling: Cavity volume: 10,000 m^3; filling rate 500 m^3/yr; filling time 5 yrs and 10 yrs, respectively; type of filling instantaneous; filling factor 25 vol. °/o and 50 vol. °/o, respectively.

Moreover, the following conservative assumptions were made:

- The cavity is a completely sealed system (no ventilation of the cavity).
- All the radiolytic gases formed by radiolysis of bitumen are released from the waste products into the atmosphere of the cavity.

To calculate the cumulative amounts of hydrogen, which is the principal component of radiolytic gases (besides hydrogen CH_4 and C_2H_x are formed reaching about 15 °/o of the hydrogen volume), the integral absorbed radiation dose of bitumen/$NaNO_3$ mixtures during storage in the prototype cavity was calculated with a "SPALT" integration program [1] as a function of the storage time for different specific activities of the products and different ages of the fission products. The program allows to calculate the inventory of fission products and the activity as a function

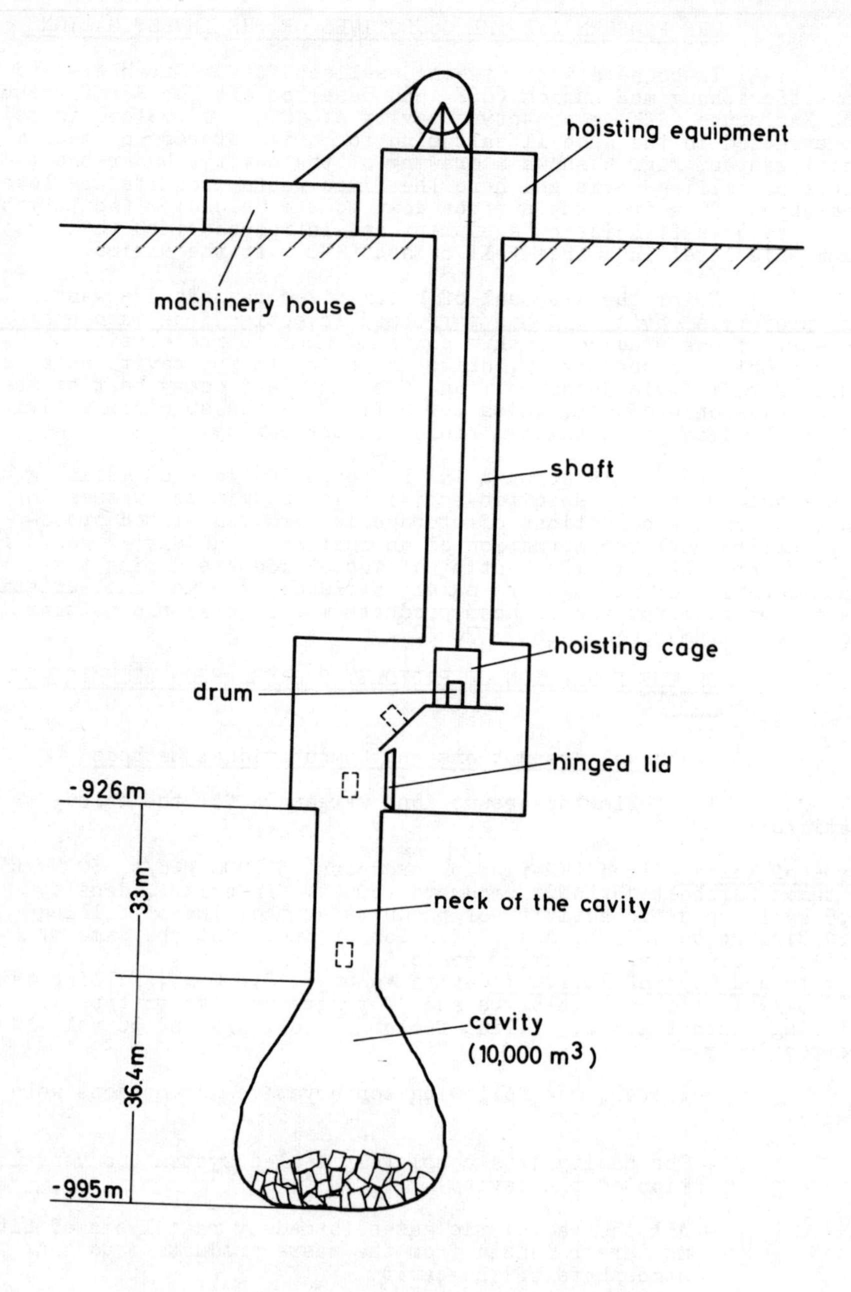

Fig. 1 Sketch of prototype cavity

of the irradiation and cooling times. The equations are solved numerically. Total absorption of ß and γ radiation was assumed in the calculations. The yields of fission products from ^{235}U fission by thermal neutrons (14 MeV) were taken from Meek and Rider [2], the half lives and the ß and γ energies of isotopes from Lederer et al. [3] and the activation cross sections from BNL [4] and W. Seelmann-Eggebert et al. [5], respectively. The calculated values for the absorbed radiation dose together with the results of investigations by Kluger [6] on the formation of radiolytic hydrogen under irradiation of bitumen/$NaNO_3$ mixtures showing that the hydrogen formation rate is $3 \cdot 10^{-3}$ cm^3 H_2/Mrad · g were used to calculate with the computer program and as a function of storage time the accumulated hydrogen amounts present in the cavity, which are directly proportional to the absorbed radiation doses.

2.2 Results

It appears from the calculations that with increasing specific activity of the products and with growing age of fission products present in the wastes (assuming the same specific initial activity) the absorbed radiation dose and, consequently, the formation of hydrogen increase. The results of calculations of the integral absorbed radiation dose in the bitumen products and of the resulting hydrogen formation in a 175 l drum filled with bitumen products have been represented in Fig. 2 as a function of the storage time. The specific activity of the products is 0.1 Ci/l for 0.5 year, 1 year, 2 years old fission products. Fig. 3 shows the amounts of hydrogen accumulated in the cavity during storage of bitumen products as a function of the storage time for a filling factor of the cavity of 50 vol. °/o (≙ 5,000 m^3 of waste). The specific activity of products is 0.1 Ci/l for 0.5 year, 1 year and 2 years old fission products. It is evident from Fig. 3 that the hydrogen formed by radiolysis of bitumen/$NaNO_3$ mixtures having a specific product activity of 0.1 Ci/l and with the fixed fission products up to 2 years of age will not give rise to an ignitable radiolytic gas/air mixture in the prototype cavity.

The limit values derived for the specific activity, taking into account all radiolytic gases (H_2, CH_4, C_2H_x) and the absorbed dose in the bitumen products in which, according to the assumptions made here, ignitable radiolytic gas/air mixtures are not formed during the entire period of disposal in the prototype cavity, have been indicated in Table I as a function of the age of fission products in the wastes and of the filling factor of the cavity.

Table I: Limit values for the specific activity and the integral absorbed radiation dose in the bitumen/$NaNO_3$ mixtures (50 wt. °/o salts) corresponding to 4 vol. °/o H_2.

Age of Fission Products [yrs]	Cavity Filling Factor [vol. °/o]	Limit Values of Specific Activity [Ci/l]	Limit Values of Absorbed Dose until Total Decay [rad]
0.5	25	0.78	$2.6 \cdot 10^7$
1	25	0.43	$2.6 \cdot 10^7$
2	25	0.26	$2.6 \cdot 10^7$
0.5	50	0.26	$8.6 \cdot 10^6$
1	50	0.14	$8.6 \cdot 10^6$
2	50	0.09	$8.6 \cdot 10^6$

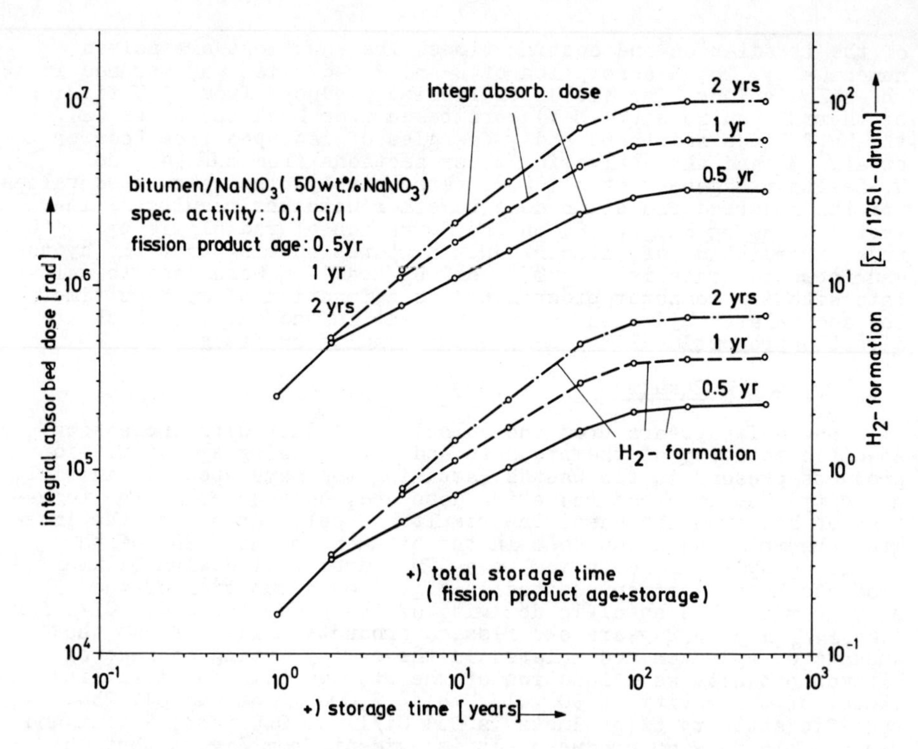

Fig. 2 Integral absorbed dose in bitumen/$NaNO_3$ mixtures and hydrogen formation in a 175 l drum filled with bitumen products

3. ON THE HEAT GENERATION FROM BITUMINIZED WASTES

3.1 Model Assumptions and Methods of Calculation

The following model assumptions were made for the investigations:

Type of Waste: Fixed LWR-fission products (33,000 MWd/t, 30 MW/t); bitumen Mexphalt 15/$NaNO_3$ products (50 wt. °/o salts), density 1.5 kg/l; specific activity of wastes at the time of filling 0.1 Ci/l up to 1 Ci/l; age of fission products (at the time of filling) 0.5 year, 1 year and 2 years.
Cavity: Diameter 22.8 m, height 36 m (corresponding to the maximum dimension of the cavity); compact filling (up to the respective level); heat conductivity of filling 0.31 W/m°C (corresponding to the waste product leaving aside the sheet metal drums); no air gap provided between the filling and the cavity wall; initial temperature of the surrounding salt 37°C.
Type of Filling: Instantaneous or step by step in constant single steps extended over 10 years.

Three different non-steady-state heat conduction programs were used for the temperature calculations. The case of instantaneous filling was first studied using an analytical formula for heat propagation based on the "WÄRMELEIT" computer program [7] with the cavity geometry approximated by a rectangular parallelepiped having the same cross section. The temperature at the cavity wall was assumed to be constant. More precise calculations of this case by means of a numerical two-dimensional heat conduction program in the "TEFELD" cylinder geometry [8] showed that the

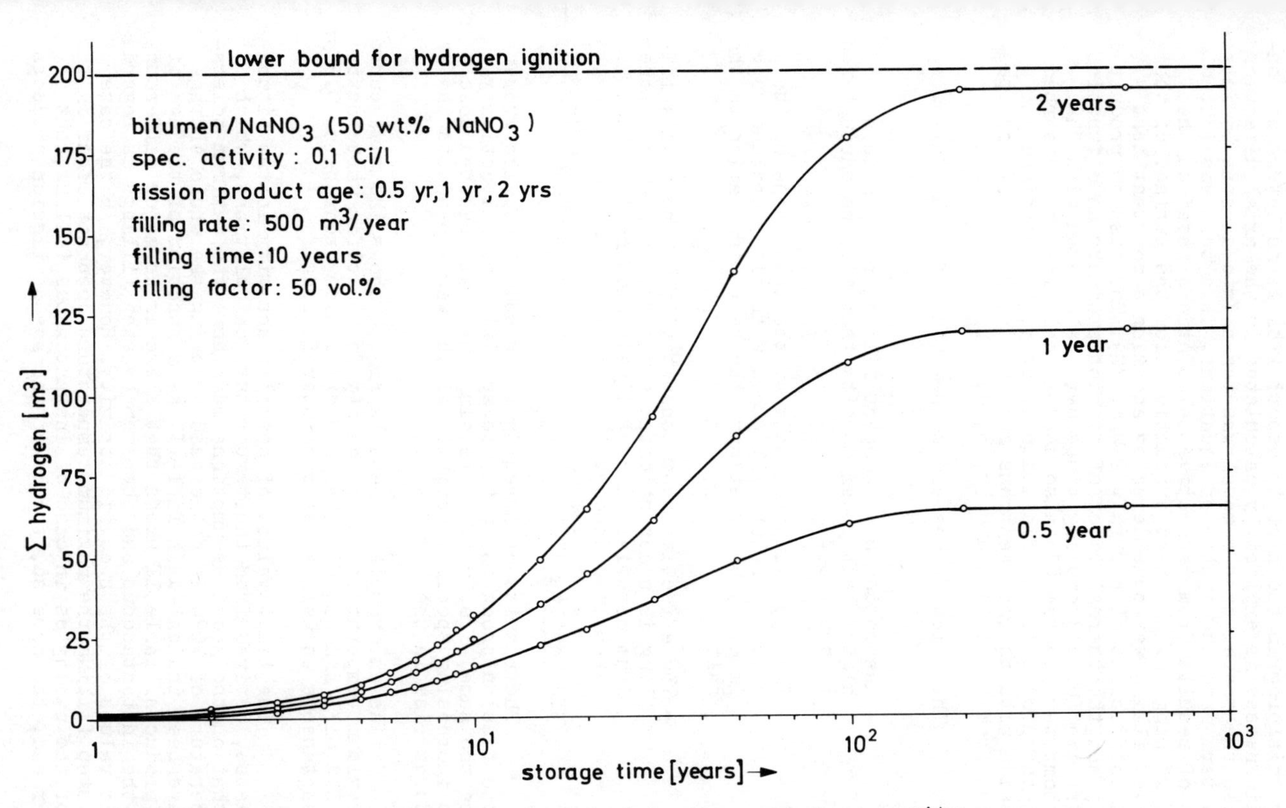

Fig. 3 Amounts of hydrogen accumulated in the prototype cavity

calculations based on the first computer program furnished maximum temperatures which were too low by only about 5 $^{o}/o$. For the case of step by step filling comprising a greater number of individual steps a special program was developed [9] allowing to simulate this process, thus simulating also continuous filling in an approximation. By this program the heat transport from plane layers of constant thicknesses is explicitly calculated in the axial direction only while the radial heat removal was taken into account by introducing a term into the system of equations. This term was fitted by means of results from respective calculations based on the "WÄRMELEIT" program. The heat evacuation from the surface of the respective filling was considered by assuming a constant heat transfer coefficient of air (α = 2 W/m^{2} ^{o}C). The fission product inventory and the thermal power of fission products as a function of the irradiation and cooling times were calculated with the "SPALT" computer program [1] also used to calculate the absorbed doses in the products. The fission product yields obtained in ^{235}U fission with thermal neutrons and the decay data of isotopes were taken from the literature [2, 3, 4, 5].

The calculations are conservative for the following reasons:

- Assumption of a compact filling results in a higher heat power density than will be encountered in practice.
- The approximation used of the true shape of the cavity (prolate ellipsoid with circular cone superimposed) underestimate the heat dissipation to the salt.
- The effective heat conductivity of the cavity filling is underestimated since the influence of the drum material on heat dissipation has not been taken into account.

3.2 Results

The results of investigations show that with growing age of the fission products (with the specific initial activity remaining the same) i.e. with increasing integral absorbed dose and with increasing specific activity of the waste products the temperatures rise in the cavity.

As an example of results Fig. 4 shows the time curve of the maximum temperature in the cavity for specific waste activities of 0.1 Ci/l and 0.3 Ci/l, an age of fission products of 0.5 yr and instantaneous as well as step by step filling.

The limit values of specific activity in the bituminized wastes, derived from the temperature calculations, which ensure that on the model assumptions made here the maximum tolerable temperature of 70°C is not exceeded during the whole storage time, have been indicated in Table II as a function of the age of fission products. Table II shows that there is no major difference between the instantaneous and the step by step fillings as regards the limit values of the specific activity. However, in the case of step by step filling the maximum temperature does not occur in the center of the cavity as in case of instantaneous filling, but at the upper end, the more so, the lower the age of fission products is.

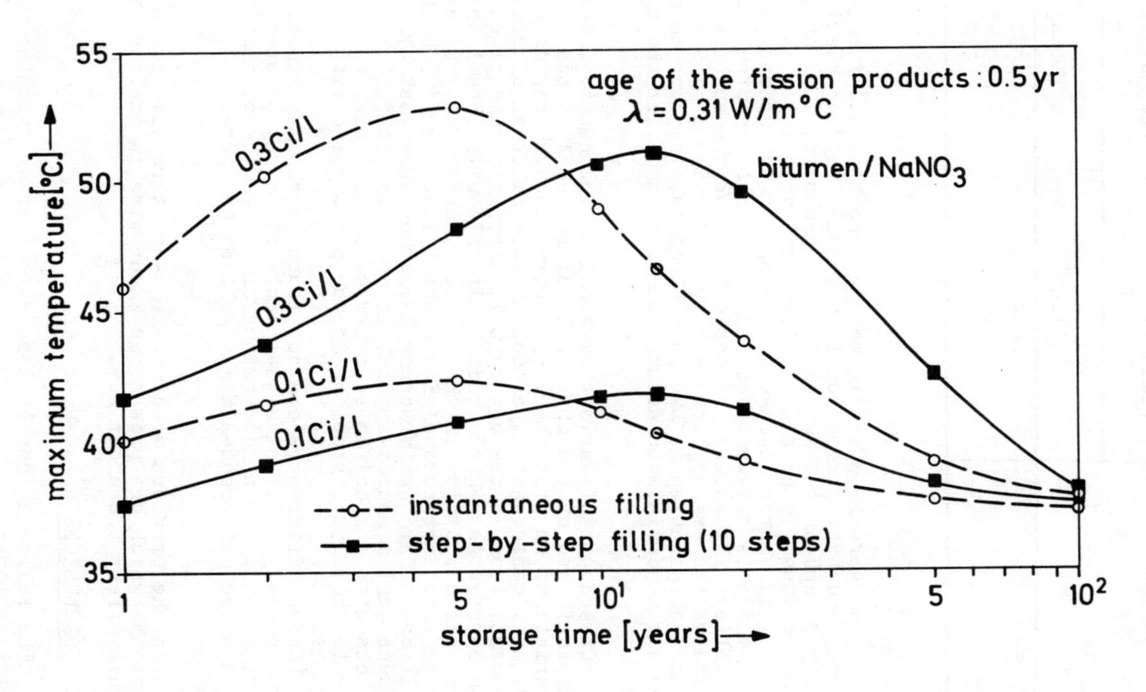

Fig. 4 Maximum temperature in the prototype cavity as a function of storage time

Table II: Limit values for the specific activity in the bitumen/$NaNO_3$ mixtures (50 wt. °/o salts) corresponding to a maximum temperature of 70°C.

age of fission products [yrs]	limit values of specific activity [Ci/l]	
	instantaneous filling	step by step filling
0.5	0.6	0.7
1	0.45	0.45
2	0.35	0.3

4. DISCUSSION

It appears from the investigations that the formation of hydrogen and heat generation from fission product bearing wastes strongly depend on the age of fission products, i.e. on the integral absorbed dose. To be able to fix limit values for the specific activity in wastes in case of storage in the prototype cavity the knowledge is necessary of the age of fission products at the time of filling. The values indicated here for the specific activity in the wastes are applicable only to fission products fixed in bitumen originating in LWR fuel elements. They are conservative values since pessimistic assumptions were made for the calculations.

On account of various external influences (e.g. ventilation of the shaft and of the upper half of the neck of the cavity as well as variations in atmospheric pressure) an exchange of air in the cavity and, consequently, a considerable dilution of radiolytic gas concentration in the cavity will certainly take place in the prototype cavity. As regards radiolysis this will entail an increase in specific activity in the wastes. Theoretical studies performed by the Gas Institute of Karlsruhe University [10] on behalf of Gesellschaft für Kernforschung Karlsruhe show that by appropriate technical measures applied in addition the filling of bitumen products with a specific activity of up to 1 Ci/l would not give rise to radiolytic problems in the prototype cavity.

Major differences are not found in the maximum temperatures of the prototype filled either instantaneously or step by step. However, with different dimensions of the cavity or different waste properties the differences might well be on larger scale.

In the previous studies the α-emitters have not been taken into account which are present as impurities in the wastes. Studies to this effect are under way.

5. CONCLUSIONS

The following conclusions can be drawn from the results of investigations:

- Disposal of medium level bitumen products generated now and over the next 3 to 4 years at the Karlsruhe Nuclear Research Center with a medium specific activity of about 0.1 Ci/l (maximum value 0.3 Ci/l) and an age of fixed radionuclides of about 0.5 year would not cause radiolytic and heat problems in the prototype cavity.

- When bitumen products shall be disposed of in non-ventilated storage spaces, an upper limit must be set for the specific activity in the waste products because of radiolytic gas formation and heat generation. This upper limit would be below the value of 1 Ci/l which under the aspect of radiation resistance of the bitumen fixing material would still be admissible, since radiolysis and heat problems arise before the bitumen is damaged by radiation.

- Disposal in subsoil storage spaces (in this case in the prototype cavity) of bitumen products having specific activities of up to 1 Ci/l is possible without hazard to the safety by radiolytic gas or heat generation, provided that the following measures are taken:

  a) Ensuring the exchange of air, which reduces the radiolytic gas concentration in the storage spaces, or filling the storage space with cover gas (e.g. CO_2, N_2).

  b) Construction of slim cavities (diameter $\ll$ height) for better radial heat dissipation to the surrounding salt or increase of effective heat conductivity of the cavity content, e.g. by providing media with a good thermal conductivity. Another possibility is offered by the interim storage of waste products prior to filling them into the prototype cavity.

REFERENCES

[1] W. Comper, Internal ASS Report 12 (1972)

[2] M. E. Meek, B.F. Rider, NEDO-12154 (1972)

[3] C. M. Lederer, I. M. Hollander, S. Perlmann, Table of Isotopes, 6th ed. John Wiley and Sons, New York (1976)

[4] BNL - 325 (1972)

[5] W. Seelmann-Eggebert, G. Pfennig, H. Münzel, Table of Nuklides (1974)

[6] W. Kluger in H. Krause, KFK 1830 (1973)

[7] E. Korthaus, unpublished results

[8] H. Schmidt, Thesis, Rheinisch-Westfälische Technische Hochschule Aachen (1971)

[9] E. Korthaus, unpublished results

[10] Engler-Bunte-Institut der Universität Karlsruhe "Gutachtliche Stellungnahme zu gastechnischen Fragen bei der Lagerung von radioaktiven Abfällen in der Prototyp-Kavernenanlage im Salzbergwerk Asse"

Discussion

Y. SOUSSELIER, France

In your calculations, did you assume that drums are destroyed by their fall into the cavity and that as a result of creeping bitumen the cavity will completely fill up or that the drums remain intact and, in this case, how do you take this into account in the thermal calculations ?

E. SMAILOS, F.R. of Germany

The first case is true. We assume destruction of the drums and complete filling of the cavity with bitumen. We did not take into account that the drum material might lead to better heat dissipation.

P.W. KNUTTI, Switzerland

We have simultaneous production of heat and H_2 during time. Which point is reached first, max. Temp. or max. H_2 production ?

E. SMAILOS, F.R. of Germany

Under the assumptions stated in the paper, the maximum temperatures cited are reached first, whereas the radiolytic hydrogen formation is a very slow process.

P.W. KNUTTI, Switzerland

Does your thermal model assume a two-dimensional heat flux , that is radial and axial ?

E. SMAILOS, F.R. of Germany

Yes, both axial and radial heat dissipation has been taken into account using a two-dimensional non-steady state heat conduction programme.

Discussion

Y. SOUSSELIER, France

Dans vos calculs, retenez-vous comme hypothèse que les fûts sont détruits pendant la chute et que par suite du fluage le bitume remplit complètement la cavité ou bien que les fûts restent intacts et dans ce cas, comment en tenez-vous compte dans les calculs thermiques ?

E. SMAILOS, R.F. d'Allemagne

Votre première hypothèse est exacte. Nous supposons que les fûts sont détruits et que la cavité est complètement remplie de bitume. Nous n'avons pas tenu compte du fait que les matériaux constituant les fûts pourraient conduire à une meilleure dissipation de la chaleur.

P.W. KNUTTI, Suisse

Nous avons une production simultanée de chaleur et de H_2 avec le temps. Quel est celui de ces deux facteurs qui est obtenu le premier : la température maximale ou la production maximale de H_2 ?

E. SMAILOS, R.F. d'Allemagne

Selon les hypothèses énoncées dans la communication, les températures maximales signalées sont atteintes en premier lieu, alors que le processus de formation d'hydrogène de radiolyse est très lent.

P.W. KNUTTI, Suisse

Votre modèle thermique suppose-t-il un flux de chaleur à deux dimensions, c'est-à-dire radial et axial ?

E. SMAILOS, R.F. d'Allemagne

En effet, nous avons tenu compte à la fois de la dissipation de chaleur axiale et radiale en utilisant un programme à deux dimensions de conduction de chaleur à l'état non stationnaire.

REVIEW OF THE RESEARCH AND DEVELOPMENTAL WORK AND EXPERIENCES ON THE BITUMINIZATION OF RADIOACTIVE WASTES IN JAPAN

T. Segawa
Tokai Works, Power Reactor and Nuclear Fuel Development Corporation
Tokai-mura, Ibaraki-ken, Japan
S. Kadoya
Ebara Manufacturing Co., Ltd.
Asahichō, Haneda, Ōta-ku, Tokyo, Japan
A. Matsumoto
Japan Atomic Energy Research Institute
Ōarai, Ibaraki-ken, Japan

Abstract

Efforts of development and improvement of the bituminization of radioactive wastes have been made by a few groups for the past ten and more years in Japan.

A new approach to incorporation of radioactive sludges into emulsified asphalts by addition of a surfactant has been studied and the leachability of solidified products has been examined.

Some developmental work of the bituminization of the evaporator concentrates expected from a reprocessing plant has been made and evaluation tests on the solidified products have been carried out.

Incorporation into bitumen of the radioactive wastes arising at the JMTR has been practiced for three years.

1. INTRODUCTION

Bituminization is a preferable solidification process of radioactive wastes in view of reducing the volume of the solidified products as well as the release of radioactivity into the environment.

Several efforts of development and improvement of theis process have been made by a few groups mainly under governmental contracts since 1966, aiming at practical use of the process at the nuclear power stations and the reprocessing plant in Japan.

The report on these efforts is presented in the following, divided into sections: 1) Development of bituminization processes for radioactive sludges and the data on evaluation of the solidified products, 2) Development of bituminization of radioactive concentrates and evaluation of the solidified products, and 3) Some operational experiences and prospects of the bituminization of radioactive wastes in Japan.

2. DEVELOPMENT OF BITUMINIZATION PROCESSES FOR RADIOACTIVE SLUDGES AND THE DATA ON EVALUATION OF THE SOLIDIFIED PRODUCTS*

2.1 Basic research /1/

In Japan, a basic research and developmental work for bituminization of radioactive wastes was initiated in 1966 by EBARA MFG. CO., Ltd. under sponsorship of Japanese government. The principal objectives of this program were to find out whether the asphalt of domestic supply could yield products of good quality and to evaluate how much radioactivity would be leached out into the surrounding water of the product when dumped into sea.

Straight, blown, and emulsified asphalts were tested with satisfactory results except cationic emulsified asphalt which requires careful operation because of foaming when radioactive wastes are intorduced.

The main features of the tests are as follows:

Sludges:

The following synthetic sludges were used for mixing in bitumen.
Copper ferrocyanide sludge: $CuSO_4 \cdot 5H_2O$, $K_4Fe(CN)_6 \cdot 3H_2O$
Ferric hydroxide calcium phosphate sludge:
$Na_3PO_4 \cdot 12H_2O$, $FeSO_4 \cdot 7H_2O$, $Ca(OH)_2$, NaOH
Aluminum hydroxide calcium phosphate sludge:
$Na_3PO_4 \cdot 12H_2O$, $Al_2(SO\ )_3 \cdot 18H_2O$, $Ca(OH)_2$, NaOH

Tracers:

^{89}Sr, ^{137}Cs spiked in each sludge before mixed with bitumen. The specific activity ranged between 2 and 10 μCi/g of dried sludge.

Equipment:

Two types of mixer with electric heater ——→ a vertical batchwise mixing tank and a horizontal continuous screw kneader.

Specimens for leaching test:

Pellets of 4 cm in diameter and 4 cm in height with surface area of 75—76 cm^2.
Measurements of leached out radioactivity were made with simulated sea water and laboratory demineralized water mostly by static test, but also by dynamic test with the specimens in the acrylic columns in which water was recirculating at a constant rate.

---

* By EBARA, Kadoya, S., Nomi, M., Hayashi, T., Sano, K., Sugimoto, S.

Test results of approximately 100days (Table I and Figures 1 and 2):

To compare the data obtained with the figures which had appeared in the published papers so far, the following classical definition of the leaching rate was used.

$$\text{Leaching rate} = \frac{C/Co}{A/W} \; (g/cm^2 \cdot day)$$

where

C : total amount of radioactivity leached per day (CPS/day)
Co : total radioactivity in test specimen (CPS)
A : surface area of test specimen (CM^2)
W : weight of test specimen (G)

Some of the findings:

a) The emulsified asphalt mixes were found better, i.e. lower in the leaching rate in sea water than the straight bitumen mixes and the increase in the solid content in the mix tended to cause an increase of the leaching rate as a fraction of the sludge that was barely coated or exposed was expected to increase.

b) As for the difference between the demineralized water and the sea water, it was found in these experiments that more activity was released in the demineralized water as suspected.

c) The tests of the bitumens of domestic supply revealed that both of the emulsified and straight bitumens were quite acceptable, in particular the straight bitumen of the needle penetration of 40/60 and the anionic emulsified asphalt were satisfactory in view of the specific weight and the leaching rate of the product.

d) To increase the solid content as high as possible, it was felt vital that the better mixing devices should be provided as the imperfect mixture with a fraction of solids left uncoated might release an excessive amount of radioactivity into water. As a good mixing was governed by proper power input for the agitator and the temperature control which was related to the viscosity of the molten mixture, the better knowledge of the physical properties of the bitumen mix was really desired.

2.2 Bituminization with emulsified asphalts /2/

A study was made on the bituminization process with emulsified asphalts. Prepared simulated chemical sludges were first mixed with base bitumen and then added with surfactant to get emulsified. After salting-out took place by addition of a proper emulsified bitumen to separate water, further mixing with heating yielded a homogeneous solidified product. With this process, favorable heat economy is well noted because of the preliminary dewatering of substantial amount of water by salting-out, which eliminates much of heating operation.

Another process of this type of bituminization with emulsified asphalt was also studied; a process which, without the use of base bitumen, starts with mixing sludes with emulsified bitumen and then goes to the addition of the emulsified bitumen of different type to separate water contained in sludges through salting-out. This process is based on the salting-out effect which is brought out at the isoelectric point to be reached by mixing cationic with anionic surfactants.

The conditioned emulsified asphalt with surfactant was charged and mixed with sludge in batchwise in a basket type centrifuge with rotated first at approximately 50 rpm and mixed the contents by built-in paddles. Rotation of the machine continued at the low revolution until most of the salted-out water was driven off out of the bulk of mixture. The revolution of the basket was then increased up to 3,000 rpm to dewater to dryness, and finally after spinning, the mixture was scraped off by the blade. As the mixture still contained water of 20 to 30 % of the total weight, it was charged into a tilted conical kneader rotating at some 45 rpm and was heated up to 130°C to evaporate the rest of the water to yield a final product.

Materials used in preliminary test:

Bitumen: a) Straight bitumen: needle penetration of 80/100
b) Cationic emulsified bitumen: MK-2 } water content: approx. 40 %
c) Anionic emulsified bitumen: MA-2 }

Sludge:
Ferric hydroxide: $Fe(OH)_2 + Ca_3(PO_4)_2$, pH = 11 , water content: 62 to 65 %
Aluminum hydroxide: $Al(OH)_3 + Ca_3(PO_4)_2$, pH = 6 to 7, water content: 68 to 73 %
Copper ferrocyanide: $Cu_2Fe(CN)_6$, pH = 7 to 8, water content: 69 to 71 %

Surfactant: Cationic aliphatic diamine

Surfactant was added to bitumen before mixing with sludges. Its dosage was 5 to 20 % of dry sludge by weight.

Cases of combination of bitumen and surfactant:

Case 1 : The cationic bitumen was used as a base bitumen and mixed with the cationic surfactant and was finally neutralized by the anionic emulsified bitumen.

Case 2 : A base bitumen was not used and the cationic surfactant was neutralized by addition of the anionic emulsified bitumen.

Case 3 : The cut-back bitumen was used as a base bitumen and mixed with the cationic surfactant and was neutralized by the anionic emulsified bituemn.

The result was summarized in Table II.

Dewatering ratio was defined as follows:

$$\text{Dewatering ratio (\%)} = \frac{\text{Amount of water salted out} \times 100}{\text{Amount of water in emulsified asphalt plus sludge}}$$

No relationship could be found between the dewatering ratio and the sludge concentration in the range of 30 to 50 %.

Following the preliminary test, the dewatering ratios and leached-out radioactivity of the products were measured.

Materials used in hot test:

Base bitumen: Straight 80/100
Neutralizing bitumen: Anionic emulsified bitumen MA-2
Sludges: $Fe(OH)_3 + Ca_3(PO_4)_2$
Tracers: ^{90}Sr, ^{137}Cs, ^{144}Ce

Results (Table III, Figures 3 and 4.):

The counting results of ^{144}Ce were discarded as they fluctuated from day to day while those of ^{90}Sr and ^{137}Cs gave the relatively consistent results which are shown in Figs. 3 and 4.

Fig. 3 shows that the release mechanism of radioactivity of the salted-out and dewatered sludge mixture may differ from that of the salted-out and heated sludge mixture although there is an uncertainty left due to the fluctuation of the counting results.

It was confirmed that both of the straight asphalt and the emulsified asphalt were workable with the process of surfactant addition to the sludge asphalt mixture and that further, the effectiveness of their use was demonstrated on their functions of the base asphalt and the neutralizing asphalt respectively to achieve the electrical neutrality for the best product.

The average dewatering efficiency was about 60 % of the total amount of the water in the salting-out sludge-bitumen mixture though the best figures obtained reached 77 %. The cationic diamine surfactant worked very efficiently and the recommended dosage was about 15 % of the dry weight of the added sludge in view of the dewatering efficiency as far as the experimental data covered.

The water expelled from the mixture contained a certain amount of radioactivity; ^{144}Ce gave the least release of the radioactivity in the separated water and ^{90}Sr came next and ^{137}Cs the last. As the sludge content increased, an increase in the radioactive concentration of the separated water was observed with the same sludge and the same nuclide.

The leaching test showed that ^{144}Ce gave the least release of the activity into water, ^{90}Sr the second and ^{137}Cs the highest in consideration of the specific activity of the mixture. As far as the result indicated, the activity concentration due to the release into water increased in accordance with the solid contents of 30, 40 and 50 % in this order but it seemed that well mixed products would reduce the difference.

2.3 Leaching test under high pressure /3/

In order to evaluate safety on sea dumping disposal of solidified radioactive wastes, various leaching tests were conducted under the simulated conditions of high pressure and low temperature in deep-sea water. In the following are shown part of the experimental studies on the leaching of ^{137}Cs from the bituminized products, with particular reference to those data on the cement-solidified products under the high pressures and at the low temperatures equivalent to those of a sea depth of 4000 m. A schematic flow diagram is seen in Fig. 5.

The mixing-ratio of the sludge to the solidfying material, with both methods of solidification, influences the leaching rate of the solidified product, showing a clear interrelationship. Figure 6 gives such a relationship between the sludge content and the leaching rate, based on the data of the products tested, having various contents of sodium sulphate concentrate, with ^{137}Cs added, solidified by slag-cement and by blown-asphalt. The fractions of the leached-out activities after 20 days of leaching, a period when leaching is almost stabilized, are compared.

It has been well noted from these studies that 1) the leaching rate of a nuclide incorporated in a solidified product varies with the type of solidifying material and the method of solidification and 2) the bituminized product has a far smaller leachability than the cement-solidified product and has the advantage of being smaller in volume for a given mass of waste. High pressure seems to suppress leaching-out of radioactivity in both methods of solidification.

3. DEVELOPMENT OF BITUMINIZATION OF RADIOACTIVE CONCENTRATES AND EVALUATION OF THE SOLIDIFIED PRODUCTS*

Since 1969, research and developmental work of bituminization of radioactive wastes, especially of evaporator concentrates expected from a reprocessing plant has been carried out by a PNC research group.

The prupose of this work was to find out the reliable operational conditions of the bituminization process and to study the behavior of the bituminized products during their storage and disposal from the safety point of view.

3.1 Test by bench-scale plant /4/

3.1.1 Experiments

A Hitachi-VL thin film evaporator with heating surface of 0.1 m^2 shown in Figure 7 was used for bituminization experiments.

A circulation system of thermofluid was employed for heating of the evaporator, bitumen melter, bitumen storage tank and tibumen feed-pipe line.

---

* By PNC, Miyao, H., Mizuno, R., Yamamoto, M.

Simulated evaporator concentrates:

The expected composition of the evaporator concentrates from the PNC reprocessing plant under test operation is as follows:

Volume	0.5 m^3/d on average
Salt content	650 g/ℓ $NaNO_3$
Activity	465γ Ci/m^3 (1160β Ci/m^3)

For the test of bench-scale plant, sodium nitrate solutions (500 g $NaNO_3$/ℓ) were used as simulated evaporator concentrates.

Bitumen:

The bitumen used in the tests was of straight type 60/80.

Operating conditions:

Evaporation rate, residual water content and fluidity of the bitumen-$NaNO_3$ mixture were determined for the following parameters:

(1) Process temperature : 160° - 250°C
(2) Peripheral speed of rotor : 4.5 - 7.3 m/sec
(3) Content of sodium nitrate : 30 - 50 wt%

Radioactivity in the off-gas and distillates of bench-scale plant:

Radioactivity in the off-gas stream and distillate of the bench-scale plant was determined by using the simulated evaporator concentrate containing radioactivities of about 10^{-1}μ Ci/ml and sodium nitrate of 500 gr/l. The radioisotopes used were ^{89}Sr, ^{103}Ru and ^{141}Ce. Radioactivity in the off-gas streams was measured by means of filter paper capable of trapping about 80 % of 0.3μ DOP (Dioctylphthalate) particles.

3.1.2 Results

In the tests of bituminization of simulated evaporator concentrates, remarkable noise occurred and the torque of driving motor increased, caused by salting out of sodium nitrate on the heated surface around the inlet of concentrates. These troubles were solved by smoothing the continuous feed and lowering the temperature of vapor chamber. Besides, sodium nitrate salted out in the form of thin film on the lower part of heated surface, but a long run test over 100 hours was carried out satisfactorily. The results of tests are shown in Table IV. At process temperatures of 160°C and 180°C, pH value of condensed water was neutral, but over 200°C, pH value tended to decrease due to the reaction between bitumen and sodium nitrate.

At a process temperature of 250°C, the bitumen-$NaNO_3$ mixture blocked up the product outlet and then coking of the mixture was observed.

Evaporation rate of water was almost unchanged at the peripheral speed of rotor between 4.5 and 7.3 m/sec.

Fluidity of bitumen-$NaNO_3$ mixtures decreased with increase of sodium nitrate content. In this plant, bitumen products containing less than 45 wt% of sodium nitrate flowed down continuously from the product outlet.

The thin film evaporator has good performance to treat the evaporator concentrate expected from the PNC reprocessing plant. However, process temperature must be decided in consideration of safety of the process.

The data on radioactivity in the distillate and off-gas streams by bench-scale test are shown in Table V. Entrained bituminous oils were present in the distillates and off-gases, and choked up the filter paper used for measurement of radioactivity in off-gases.

These bituminous oils picked up radioactivity, so that distillates should be treated to remove the oily matter prior to discharge. Radioactivity in the distillates and off-gases with the bench-scale tests was similar to those at Mol and Marcoule.

3.2 Evaluation of the Solidified Products [5] [6]

3.2.1 Experiments

Irradiation test:

External irradiation tests were carried out by using ^{60}Co-source for open and sealed samples in glass tube. The open samples filled in 45 mm ϕ cylinder of pasteboard were used for the determination of swelling, change of physical properties, and leaching rate. The sealed samples, 10g of bituminized products in vacuum, were used for the determination of gas gegerated by radiolysis of asphalt. The composition of the gas was analysed by mass spectrographic method.

Burning tests:

Since a main chemical component of the evaporator concentrates is sodium nitrate, $NaNO_3$-bitumen mixture burns very vigorously at above 320°C. Therefore, differential thermal analysis was carried out and the behavior of the mixtures in a dry oven at a constant temperature was observed in detail.

Leachability:

Leaching tests were carried out on the bitumen products obtained by using simulated evaporator concentrates containing sodium nitrate and radioisotopes (^{90}Sr, ^{106}Ru, ^{137}Cs) in the laboratory experiments. The bitumen specimens (cylinders of 45 mm diameter and 44 mm height) contaning about 40 % of sodium nitrate and about 2 mCi of each radioisotope were prepared. The grade of bitumen used in the experiments was distillation-bitumen of penetration 60 - 80. The leaching procedure was based on the standard method of IAEA's proposal, except the temperature of the leaching tests which was maintained at bout 5°C in an electric refrigerator to simulate temperature of the deep sea and prevent the deformation of the specimen.

Hydraulic high pressure test:

To estimate the pressure effects on $NaNO_3$-bitumen mixture, the mixtures packaged with 200 ℓ drum were pressed in water up to 500 kg/cm^2 at the temperature of 10°C by the high pressure leaching test apparatus for full size waste solid shown in Figure 8.

3.2.2 Results and Discussion

By the irradiation of the samples, the swelling and the evolution of gas were caused by radiolysis. Many gas bubbles were observed in this specimen. The formation of gas bubbles as a result of radiolysis caused an increase of the volume. Figure 9 shows the volume increases of the samples of straight and blown bitumen containing 50 wt% of sodium nitrate. The volume of the generated gas is shown as a function of exposure dose in figure 10. It was about 1×10^{-2} cm^3/g·10^6 R at the exposure doses from 5×10^6 R to 1×10^8 R and there was no difference between the exposure dose rates of 1×10^6 R/h and 1×10^5 R/h. The main component of radiolysis gas was hydrogen. Small quantities of hydrocarbons and carbonmonoxide were also detected.

No influence on the characteristics (flash point, penetration, softening point, viscosity) was observed up to 1×10^8 R of exposure dose. The straight bitumen was supperior to blown bitumen as to the leachability and the flash point, and inferior as to the increase of volume by irradiation. But, since the increase of volume after irradiation at doses up to 1×10^7 R was minor, the distillation bitumen 60/80 is considered to be acceptable for bituminization of the evaporator concnetrate expected from the PNC reprocessing plant.

The results of differential thermal analysis of $NaNO_3$-bitumen mixture indicated the vigorous exothermic reaction at about 400°C. However, the tests in a dry ov at a constant temperature showed that bitumen coked gradually at about 230°C and the temperature of the sample rose remarkably at over 250°C, and at 260°C some specimen finally burned vigorously (Table VI). Therefore, bituminization process of radioactive liquid wastes containing sodium nitrate should be applied below 230°C. On the other hand, it was found that the reaction between bitumen and sodium nitrate could be detected on lowering pH of the distillates.

The leaching curves of radioactivity and sodium from the bituminized solid containing 40 % sodium nitrate over about fourteen months are shown in Figure 11. The leaching rates of sodium and strontium were almost similar and those of cesium and ruthenium were lower by 1 to 2 orders of magnitude than that of sodium. However, since the radioactivity in leaching solution was very low and hence the accuracy of measurements was low, leaching tests should be carried out on the specimen incorporated with the waste solutions of much higher activity arising at the PNC reprocessing plant.

Results of the experiments on the pressure effects to the package of 200 ℓ drum filled with $NaNO_3$-bitumen mixture are summarized as follows:

1) Top and bottom sides of the drums were dented regardless of containing ratio of $NaNO_3$ in the mixture, 2) the wall of the drum was slightly dented when the mixture had voids to some extent, and 3) no weight change of the drum was detected through the test.

Putting all accounts together the mixture packaged with drum had almost no essential defects by the simulated disposal to the deep sea bed at a depth of 5,000 meters. In order to make the disposal safer, it is desired that the voids in the mixture and the gap of the upper part of the drum should be minimized.

4. SOME OPERATIONAL EXPERIENCES* AND PROSPECTS OF THE BITUMINIZATION OF RADIOACTIVE WASTES IN JAPAN** [7]

The first bituminization was installed at the Ōarai research establishment of the JAERI in 1973 to treat chemical sludge and evaporator concentrate coming out respectively from a 10 m^3/h chemical treatment facility for low level liquid waste and a 1 m^3/h evaporator for medium level liquid waste.

The unit has been successfully operated since its first hot operation at the end of 1973.

4.1 Design and construction of bituminizing equipment

JAERI and JGC (Japan Gasoline Company) cooperated to design the unit. The following design conditions were taken into considerations for a batchwise process system.

1) Capacity: 100 l/batch of product or 50 l/h of water evaporation,
2) Materials to be treated: chemical sludge, evaporator concentrate, and if possible, incinerator ash, too,
3) High reliability and simplicity,
4) Adaptability to variety of process conditions for experimental purposes.

The process scheme was determined in reference to the Mol process, known of its successful operation for many years, and the flow diagram including a freezing and thawing unit by EBARA at a pretreatment stage of chemical sludge is shown in Figure 12.

A specially designed mixing evaporator was for the first time introduced to the bitumining process. This unit is a bathtub-shaped vessel with a jacket on the wall and equipped with two horizontal axis impellers in it. A main impeller which rotates slowly at 60 rpm provides for radial and axial mixing and is composed of annular tubes and hollow paddles through which thermo-oil is recirculated for heating. In other words, the surfaces of the shaft and paddles also serve for transferring heat besides the heating jacket of the vessel body to evaporate water. Axial mixing is further assisted by an ordinary small screw impeller installed at the bottom of the vessel as shown in Figure 13.

While the thermo-oil is heated up to the operational temperature (max. 300°C) with a propane gas furnace, our specially devised impeller enables us to mix

* By JAERI, Katsuyama, K., Nishizawa, I., Fukuda, K.
** By EBARA, Kadoya, S., Nomi, M.

and heat the highly viscous bitumen enough to get homogeneous products by applying moderately low temperature oil as heating medium.

The whole mixing evaporator unit is weighed continuously by means of load cells so as to observe the progress and the termination of the process. A U-shaped cake feeder with a transparent plastic cover on top for protection has an improvement devised to prevent cakes from sticking. The principal body of the unit is protected with radiation shieldings of concrete or iron blocks.

4.2 Operational experiences and resutls

After completion of the construction in April, 1973, the unit underwent cold (non-active) tests for confirmation of functioning of each principal component and required a few improvements based on the results of tests. The system entered into its hot (active) operation in December, 1973, and has treated the whole chemical sludges and part of evaporator concentrates generated at Ōarai site.

Before November, 1973, the unit had been used only for the chemical sludge treatment, because the sludges had already been stored since the end of 1971. In December, 1973, the evaporator concentrate was first treated by the unit, and in the two cases of concentrate runs, the concentrate was bituminized together with the chemical sludge in the same batch, because the quantity of concentrate was not sufficient to make up a full batch of bituminized product. The whole operational results are listed in Table VII. 200 liter steel drums with 5 cm inner concrete lining (effective volume: 120 liters) were used to collect the bituminized product. After cooling down of the product, the top of container was sealed off with concrete. The product packages have been stored in the outdoor storage yard of stockade style.

At the stage of cold test operation, the following troubles were encountered and all of them were settled before the hot operation:

1) Plugging by the product at the outlet valve of the mixing evaporator.
   (Solved by change of valve type and valve size up)

2) Super-heating at the product outlet line.
   (Switching of electric heater on the line to the oil trace heater)

3) Foaming in the mixing evaporator
   (Suppressed by use of a small quantity of silicon oil of emulsion type)

4) Faulty work of weighing mechnism of mixing evaporator.
   (Remedied by modification of some pipe lines fixed to the mixer which disturbed transmission of the real weight on the strain gage)

During the hot operation the followings were noticed.

1) The screw impeller was found ineffective for mixing and other purposes. It rather hindered the smooth flowing-out of the product.

2) The over-all heat transfer coefficients were observed as 100 Kcal/$m^2 \cdot h \cdot °C$ for sludge treatment and 130 Kcal/$m^2 \cdot h \cdot °C$ for concentrate treatment. These figures are remaining unchanged for more than two years of operation.

3) The light oil entrained in steam was about 2000 ppm as observed in condensate. The oil was collected and separated by an oil trapping system with trap-pot and ceramic filter.

4) Oil mist entrained in vent gas was not significantly observed. The vent gas was treated through two stages of prefilter and one stage of HEPA filter after dilution with other vent gas from the vessel. No special pressure increase was observed on these filter elements.

5) In some cases a small amount of bitumen was noticed seeping out through the axial seal mechanism (gland packing seal) of the screw impeller. This has been the most troublesome problem experienced with the unit so far. Deciding a complete countermeasure against this trouble will take some time. Meanwhile, a

few catching pots are being used at present to collect the seeping oil.

4.3 Prospects of the bituminization of radioactive wastes in Japan

In Japan, it has been a practice to solidify radioactive wastes arising at nuclear power stations with cement so far. However, the advantages of bituminization of considerable reduction of volume of the solidified products and of especially marked reduction of leachability of radioactivity have been recently attracting attention to adoption of bituminization process of wastes.

This led to the decision by PNC in 1973 to employ this process at the "FUGEN", an advanced thermal reactor, then to another by Japan Atomic Power Company for solidifying evaporator concentrates at the Tsuruga Nuclear Station (BWR of 350 MWe), both at Tsuruga site. In 1975, again this process was adopted by KANSAI Electric Power Company for its Mihama Station (PWR Reactors Nos. 1, 2, and 3 of 1650 MWe in total) to solidify boric acid waste concentrate. The contracts of construction of all these three plants were awarded to EBARA MFG. CO., Ltd. which has a technical licence agreement with Saint-Gobain Techniques Nouvelles for the French CEA Process. Further adoption of the process at some more places is expected in succession in the very near future.

In order to establish proper and well fitted operational guides for the bituminizing equipment prior to the start-up at these power stations, one complete set of equipment comprised of a wiped film evaporator LUWA NL-150 of 50 Kg/h has already been installed at EBARA's factory and is expected to start mock-up tests quite soon.

For the reprocessing plant of PNC at Tokai, now under various preoperational tests, the construction program of an inbitumining facility to be integrated has been started and the completion of the plant will be expected in three or four years.

Literature

(1) Kadoya, S., Nomi, M., Sano, K., Hayashi, T., The bench-scale experiments on the sludge incorporation into the emulsified and straight asphalts. Research Co-ordination Meeting on the Incorporation of Radioactive Wastes in Bitumen, Dubna, 9-13 Dec. 1968.
(2) Kadoya, S., Sugimoto, S., Shimogori, M., Hayashi, T., The sludge incorporation into the emulsified asphalt by addition of a surfactant. Same as in (1).
(3) Kadoya, S., Sugimoto, S., Hayashi, T., et al., Protecting the ocean and atmosphere from contamination by radioactive waste disposal activities. Proc. Vol. 11, p. 308, 312, 322 of Geneva Conf. by IAEA, 1972.
(4) Segawa, T., Yamamoto, M., Miyao, H., et al., Development of bituminization process for radioactive wastes, test by bench-scale plant. Tokai Works Semi Annual Report PNCT831-73-01, p. 39-45, 1973.
(5) Segawa, T., Yamamoto, M., Miyao, H., et al., Development of bituminization process for radioactive wastes, evaluation of bitumen product. Tokai Works Semi Annual Report PNCT831-73-01, p. 45-51, 1973.
(6) Miyao, H., Mizuno, R., Muto, H., et al., Some safety data of bituminization process for radioactive wastes. Tokai Works Semi Annual Report PNCT831-74-02, p. 101-106, 1974.
(7) Nishizawa, I., Fukuda, K., Operational results of bituminization unit. Health Physics and Safety in JAERI No. 16, p. 157-161, 1974.

TABLE I-a. RESULTS OF LEACHING TEST OF BITUMEN MIXES IN WATER.

No.	Asphalt		Sludge		Mix'g ratio w/o dry solid	Tracer		Test Piece				Leached activity cpm/ml	Leaching rate g/cm day			Static or dynamic
	Make	Weight g-dry	Make	Weight g-dry		Nuclide	μCi/g-dry solid μCi/piece	Weight g	Vol cm^3	Sp. Wt.	Surface cm^2				Vol ml	
1	Emulsified[MA]	200	K-Cu	86	30	^{137}Cs	6.9 227	110	89	1.29	123	25	8.6×10^{-7}	Sea	900	S
2	Emulsified[MA]	100	K-Cu	66	40	^{137}Cs	6.9 350	127	94	1.35	140	55	1.29×10^{-6}	Sea	900	S
3	Emulsified[MA]	100	K-Cu	100	50	^{137}Cs	6.9 496	144	94	1.53	137	80	1.27×10^{-6}	Sea	900	S
4	Straight	100	K-Cu	43	30	^{137}Cs	6.9 144	70	53	1.32	78	30	1.8×10^{-6}	Sea	1000	S
5	Straight	100	K-Cu	67	40	^{137}Cs	6.9 182	66	52	1.20	73	270	1.35×10^{-5}	Sea	1000	S
6	Straight	100	K-Cu	67	40	^{137}Cs	10.3 268	65	50	1.30	76	108	3.47×10^{-6}	Sea	1000	S
7	Straight	200	K-Cu	133	40	^{137}Cs	3.5 87	62	47	1.32	72	120	1.20×10^{-5}	Sea	1000	S
8	Straight	100	K-Cu	100	50	^{137}Cs	6.9 248	72	56	1.29	77	215	8.0×10^{-6}	Sea	1000	S
9	Straight	100	Fe	67	40	^{137}Cs	10.3 453	110	78	1.42	88	1500	3.76×10^{-5}	Sea	900	S
10	Straight	150	K-Cu	64	30	^{137}Cs	6.9 149	72	56	1.28	81	25	1.43×10^{-7}	demi	1000	S
11	Straight	100	K-Cu	67	40	^{137}Cs	10.3 280	68	54	1.26	78	70	2.16×10^{-6}	demi	1000	S
12	Straight	100	K-Cu	100	50	^{137}Cs	6.9 276	80	56	1.43	80	205	7.46×10^{-6}	demi	1000	S
13	Straight	200	K-Cu	133	40	^{137}Cs	10.3 506	123	92	1.34	730	170	5.75×10^{-7}	demi	1000	D
14	Straight	100	K-Cu	43	30	^{137}Cs	3.5 122	116	93	1.25	320	10	2.70×10^{-7}	Sea	1000	D
15	Emulsified[MA]	100	Al	43	30	^{89}Sr	2.0 69	115	90	1.27	138	7	1.26×10^{-5}	Sea	900	S
16	Emulsified[MA]	100	Al	67	40	^{89}Sr	2.0 86	108	82	1.32	135	14	1.98×10^{-5}	Sea	900	S

Note: 1) Emulsified asphalt-MA: aniomic containing 40% water
2) K-Cu Sludge: $CuSO_4+K_4Fe(CN)_6$
3) Al sludge: $Na_3PO_4+Al_2(SO_4)_3+Ca(OH)_2+NaOH$
4) Fe sludge: $Na_3PO_4+FeSO_4+Ca(OH)_2+NaOH$

TABLE I-b. RESULTS OF LEACHING TEST OF BITUMEN MIXES IN WATER.

No.	Asphalt		Sludge		Mix'g ratio w/o dry solid	Tracer		Test Piece				Leached activity cpm/ml	Leaching rate g/cm^3 day			Static or dynamic
	Make	Weight g-dry	Make	Weight g-dry		Nuclide	μCi/g-dry solid μCi/piece	Weight g	Vol cm^3	Sp. Wt.	Surface cm^2				Vol ml	
17	Emulsified[MA]	200	Al	200	50	^{89}Sr	2.0 69	68	48	1.42	74	26	5.22×10^{-5}	Sea	1000	S
18	Emulsified[MA]	200	Al	200	50	^{89}Sr	2.0 78	78	53	1.47	77	18	3.55×10^{-5}	demi	1000	S
19	Straight	100	Al	43	30	^{89}Sr	2.0 37	61	50	1.22	74	20	6.66×10^{-5}	Sea	1000	S
20	Straight	100	Al	43	30	^{89}Sr	2.0 36	60	48	1.25	75	13	4.33×10^{-5}	demi	1000	S
21	Straight	100	Al	67	40	^{89}Sr	2.0 55	69	52	1.32	77	26	6.22×10^{-5}	Sea	1000	S
22	Straight	100	Al	67	40	^{89}Sr	2.0 54	68	51	1.33	74	20	5.11×10^{-5}	demi	1000	S
23	Straight	100	Al	100	50	^{89}Sr	2.0 75	75	52	1.44	78	185	3.5×10^{-4}	Sea	1000	S
24	Straight	100	Al	100	50	^{89}Sr	2.0 75	75	53	1.41	76	290	3.1×10^{-4}	demi	1000	S
25	Straight	100	Al	67	40	^{89}Sr	3.03 84	69	53	1.30	78	60	9.56×10^{-5}	Sea	1000	S
26	Straight	100	Al	67	40	^{89}Sr	3.03 85	70	52	1.34	78	50	7.67×10^{-5}	demi	1000	S
27	Straight	100	Al	67	40	^{89}Sr	2.0 58	72	56	1.28	79	25	5.93×10^{-5}	demi	1000	S
28	Straight	100	Fe	67	40	^{89}Sr	2.0	115	82	1.40	80	40	9.26×10^{-5}	Sea	900	S
29	Straight	200	Al	86	30	^{89}Sr	2.0 58	96	82	1.17	104	32	7.23×10^{-5}	Sea	900	S
30	Emulsified[MA]	100	Al	100	50	^{89}Sr	2.0 128	128	90	1.42	500	210	6.36×10^{-5}	Sea	1000	D
31	Straight	100	Al	67	40	^{89}Sr	2.0 50	62	59	1.27	190	160	1.57×10^{-4}	demi	1000	D

Note:
1) Emulsified asphalt-MA: aniomic containing 40% water
2) K-Ku sludge: $CuSO_4+K_4Fe(CN)_6$
3) Al sludge: $Na_3PO_4+Al_2(SO_4)_3+Ca(OH)_2+NaOH$
4) Fe sludge: $Na_3PO_4+FeSO_4+Ca(OH)_2+NaOH$

TABLE II. SLUDGE INCORPORATION IN ASPHALT WITH SURFACTANT

Nuclide : ^{144}Ce, ^{137}Cs, ^{90}Sr
Sludge : Mixture of ferric hydroxide and calcium phosphate
Surfactant: Cationic diamine

Nuclide	Ferric hydroxide sludge					Base asphalt		Neutralizing asphalt		Separated water by surfactant					Final product
	Water % wt	Dry wt g	Solid cont't % wt	Total activity μc	Specific activity μc/g	Type	wt g	Type	wt g	Water Vol ml	Eff. %	pH	Total activity μc	Specific activity μc/g	Weight g
^{144}Ce	65	2440	30	23912	9.8	Straight	1700	anion emuls'd	300	950	53.5	8.9	17 (0.07%)	1.8×10^{-2}	3500
	65	1600	40	15680	9.8	"	670	"	250	740	62.2	7.8	3.5 (0.02%)	4.8×10^{-3}	1900
	*71	1850	50	16355	6.3	"	600	"	450	870	55.0	6.0	71.5 (0.4%)	8.2×10^{-2}	2120
^{137}Cs	65	3000	30	6000	1.1	"	2110	"	370	1150	48.0	9.1	322 (5.4%)	2.8×10^{-1}	4200
	65	4000	40	8000	1.2	"	1680	"	625	1950	67.0	8.4	571 (7.2%)	2.7×10^{-1}	4050
	65	3700	50	7400	1.3	"	1180	"	400	1650	63.0	8.2	446 (6.0%)	2.7×10^{-1}	3340
^{90}Sr	65	1600	40	10720	6.2	"	670	"	250	570	52.5	7.8	108 (1.0%)	1.9×10^{-1}	2300

* Copper ferrocyanide sludge was used.

TABLE III. RESULT OF PRELIMINARY TEST OF THE SALTING-OUT EFFECT OF SURFACTANTS.

	Test No.	Base asphalt		Sludge			Neutralizing asphalt		Sludge content		Water separated			Penet-ration	Soft'g Point
		Type	Weight (g)	Type	Water (%)	Wt.added (g)		Weight	Asphalt Total(g)	Ratio	pH	Water (ml) vol	(%)		
Base: Cationic emulsified	1-1	Cationic emulsified	83.5	$Fe(OH)_2Ca_3(PO_4)_2$	65	122	Anionic emulsified	83.5	100	30	7.9	73	50.0	82	54.2
	2					180				40	8.2	85	44.6	44	99.6
	3					284				50	8.1	130	42.1	22	-
	4					249				60	8.1	128	37.0	-	-
	2-1	Cationic emulsified	83.5	$Cu_2Fe(ON)_6$	71	148	- ditto -	83.5	100	30	5.6	110	64.0	19	54.5
	2					230				40	6.7	143	62.2	13	97.3
	3					342				50	5.9	124	40.6	7	-
	3-1	Cationic emulsified	83.5	$Al(OH)_3Ca_3(PO_4)_2$	73	158	- ditto -	83.5	100	30	8.4	130	71.0	16.6	68.0
	2					246				40	8.4	156	65.0	84	86.8
	3					370				50	9.2	250	77.0	54	-
No base asphalt; neutralized by anionic emulsified	4-1			$Fe(OH)_2Ca_3(PO_4)_2$	65	100	- ditto -	310	186	11.9	-	85	50.0	95	49.0
	2					200		270	162	28.6	6.5	122	53.5	-	-
	3					200		170	102	38.5	9.7	125	66.0	54	80.2
	4					200		140	84	41.6	7.0	130	76.5	85	95.6
	5					100		60	36	46.6	7.8	62	72.0	23	-
	6					122		60	36	51.7	8.2	60	60.0	68	52.5
	5-1			$Al(OH)_3Ca_3(PO_4)_2$	73	200	- ditto -	400	240	17.7	7.4	140	50.5	94	-
	2					200		300	180	24.4	7.3	148	58.8	66	-
	3					200		200	120	29.4	6.9	221	56.0	57	-
	4					200		180	108	31.6	7.9	132	63.1	-	-
	5					200		160	96	34.2	7.9	137	62.3	47	62.3
Base: Straight	6-1	Straight	86	$Al(OH)_3Ca_3(PO_4)_2$	73	159	- ditto -	20	98	30	6.2	66	53.5	38	81.5
	2		71			247		35	92	40	6.9	122	67.0	29	89.0
	3		74			370		38	97	50	7.1	208	73.5	16	97.0
	4	Straight	86	$Fe(OH)_2Ca_3(PO_4)_2$	65	122	- ditto -	15	95	30	6.3	50	56.3	50	67.5
	5		80			181		10	86	40	7.5	77	54.4	46	94.1
	6		80			285		52	111	50	8.2	132	63.9	-	-
Base: Cutback	7-1	Cutback	80	$Al(OH)_3Ca_3(PO_4)_2$	73	159	- ditto -	12	87	30	7.2	65	53.0	-	-
	2		80			247		40	104	40	7.1	120	62.5	-	-
	3		80			370		80	120	50	7.4	196	62.0	-	-
	4	Cutback	80	$Fe(OH)_2Ca_3(PO_4)_2$	64	122	- ditto -	15	89	30	7.4	45	53.0	-	-
	5		90			181		20	102	40	7.5	72	57.9	-	-
	6		90			285		50	120	50	8.5	124	65.0	-	-

TABLE IV. SOME RESULTS OF THE BENCH-ACALE PLANT TESTS

Run No.	Flow rate		Product		Temperature °C			Rotor		Remarks
	Bitumen (kg/h)	Simulated waste (1/h)	Solid (%)	Water (%)	Thermo-fluid inlet	Bitumen inlet	Simulated waste inlet	Revolution (rpm)	Shape	
1	5.01	4.73	32.0	0	247.5	224	15	1380	Flat	Blocked
2	6.18	8.74	41.4	0.2	227	135	25	"	"	
3	5.70	7.60	40.0	0.3	217.5	170	25.5	"	"	
4	4.98	7.70	43.9	2.2	197	186	24	"	"	
5	4.50	5.47	37.8	2.7	177	156	20	"	"	
6	4.29	5.45	38.9	4.5	165	145	23	"	"	
7	5.01	7.55	43.0	2.3	196	189	20	"	Special	Salt out on the blades
8	5.46	7.90	42.0	0.35	217	184	25	"	Flat	
9	5.46	8.03	42.4	0.6	216	183	25	1110	"	
10	5.46	7.95	42.1	0.8	215	181	25.5	860	"	
11	9.36	7.60	28.9	0.9	216.5	185	25.5	1380	"	
12	3.78	7.76	50.9	0.1	218.5	155	25	"	"	Poor fluidity

TABLE V. RADIOACTIVITY IN THE DISTILLATES AND OFF-GAS STREAMS WITH THE BENCH-SCALE PLANT

Feed solution		Distillates		Off-gas streams	
Radioisotope added	Solution[1)]	Radioactivity ($\times 10^{-5}\mu Ci/ml$)	DF[3)]	Radioactivity[2)] ($\times 10^{-8}\mu Ci/cc$)	DF[4)]
^{89}Sr	A	6.7	3.1×10^{3}	8.4	4.9×10^{4}
	B	4.8	3.1×10^{3}	4.6	6×10^{4}
^{103}Ru	A	9.4×10^{-1}	1×10^{4}	4.9×10^{-2}	4.6×10^{5}
	B	4×10^{-1}	1.8×10^{4}	1.5×10^{-1}	1×10^{5}
^{141}Ce	A	5)			
	B	1×10^{-1}	9.4×10^{4}	1.9×10^{-1}	1.1×10^{5}

1) $NaNO_3$ (500 g/l) solution diluted with tap water (A) or demineralized water (B).

2) Radioactivity caught on the filter paper.

3) Decontamination factor of distillates $= \frac{\text{radioactivity in feed solution } (\mu Ci/ml)}{\text{radioactivity in distillates } (\mu Ci/ml)}$

4) Decontamination factor of off-gas streams $= \frac{\text{radioactivity fed } (\mu Ci)}{\text{radioactivity caught on the filter paper } (\mu Ci)}$

5) Wide variation due to formation of precipitation in feed solution.

TABLE VI. BURNING TEST: BEHAVIOR OF $NaNO_3$-BIBUMEN MIXTURES* IN A DRY OVEN UNDER THE CONSTANT TEMPERATURE

Temperature kept in a dry oven	Exothermic reaction		
	Coking,	Temperature rise (°C)	Burning
220	—	—	—
230	○ (slightly)	—	—
240	○	2	—
250	○	11	—
]60	○	41	○ began at 306°C

* Mixtures were of the blown bitumen (20/30) containing 40 wt% of $NaNO_3$.

TABLE VII. OPERATIONAL RESULTS

Sludge Run (Total: 48 Runs)	Used Bitumen	Straight Asphalt 60/80 or 80/100
	Sludge Component	$Fe(OH)_3/(Fe(OH)_3+Ca_3(PO_4)_2)$=20∿45
	Treated Volume	Av. 1.2m^3/Run, Total, Volume: 59.2m^3, Solid Contents of Sludge: About 5w/o
	Operational Condition	Thermo-oil Temperature: Max. 300°C, Operational Time: 3.5∿7.0 Hr/Run Mixing Temperature: 200 230°C, Evaporation Rates of Water: Max. 31 kg/H
	Product	Specific Gravity: Av. 1.43, Treated Volume: About 100 ℓ/Run, Radioactivity: Av. 2.0 mCi/Run Mix. Ratio; Bitumen: Solid = 57:43(Av.) Surface Dose Rates on Product: Av. 13 mR/H
Evaporator Concentrate Run (Total: 2 Runs)	Used Bitumen	Straight Asphalt 60/80
	Concentrate Component	No_3^-: 0.5%, Cl^-; 0.3%, SO_4^{2-}: 1.6%, Fe: 1600ppm
	Treated Volume	Av. 0.18m^3/Run, Total Volume: 0.36m^3, Solid Contents of Concentrate: About 8w/o
	Operational Condition	Thermo-oil Temperature: Max.300°C, Operational Time: 2.0∿3.5Hr/Run Mixing Temperature: 130∿150°C (Only Evaporator Concentrate) Evaporation Rates of Water: Max.62.5 kg/H
	Product	Specific Gravity: Av.1.34, Treated Volume: About 100 ℓ/Run, Radioactivity: Av.7.0 mCi/Run, Mix. Ratio; Bitumen: Solid=65.35(Av.) Surface Dose Rates on Product: Av.20mR/H(10 parts of 35 is solid of concentrate, 25 parts of 35 is solid of sludge.)

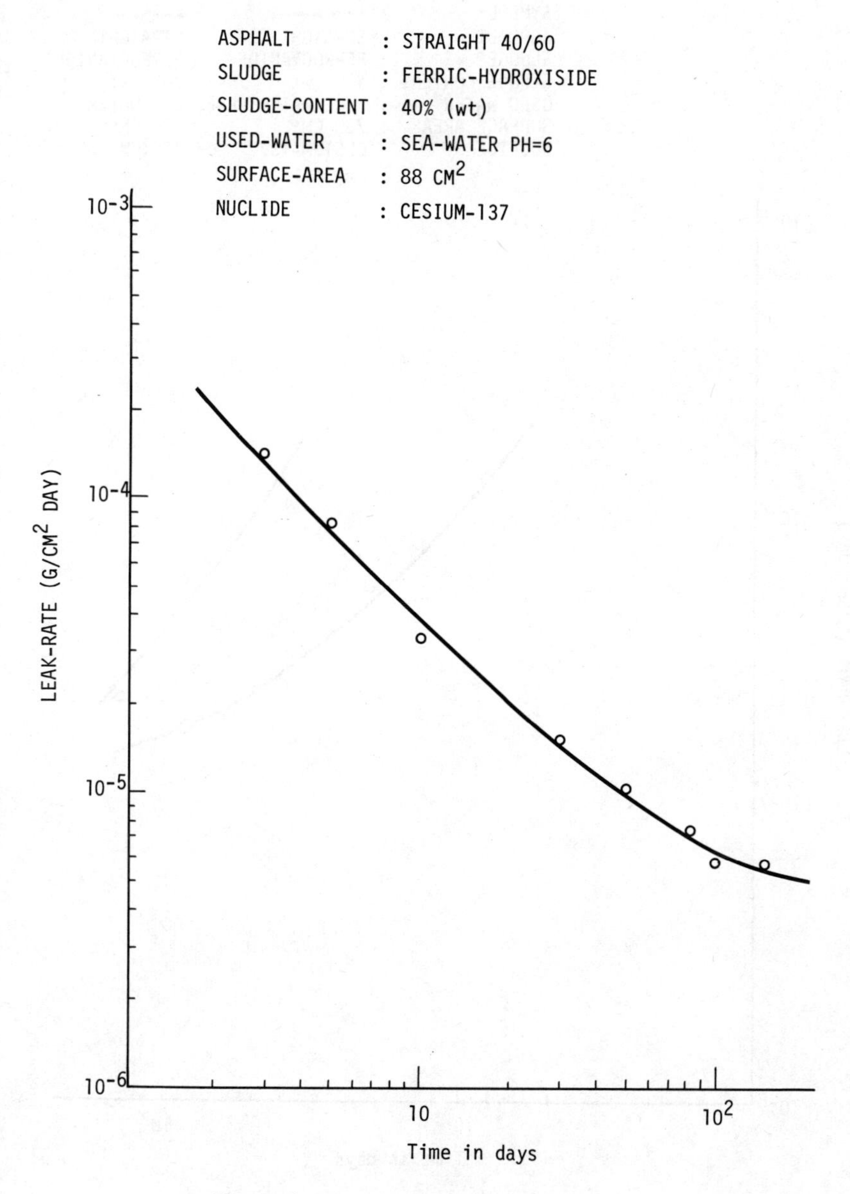

Figure 1 Leak-rate of static-test

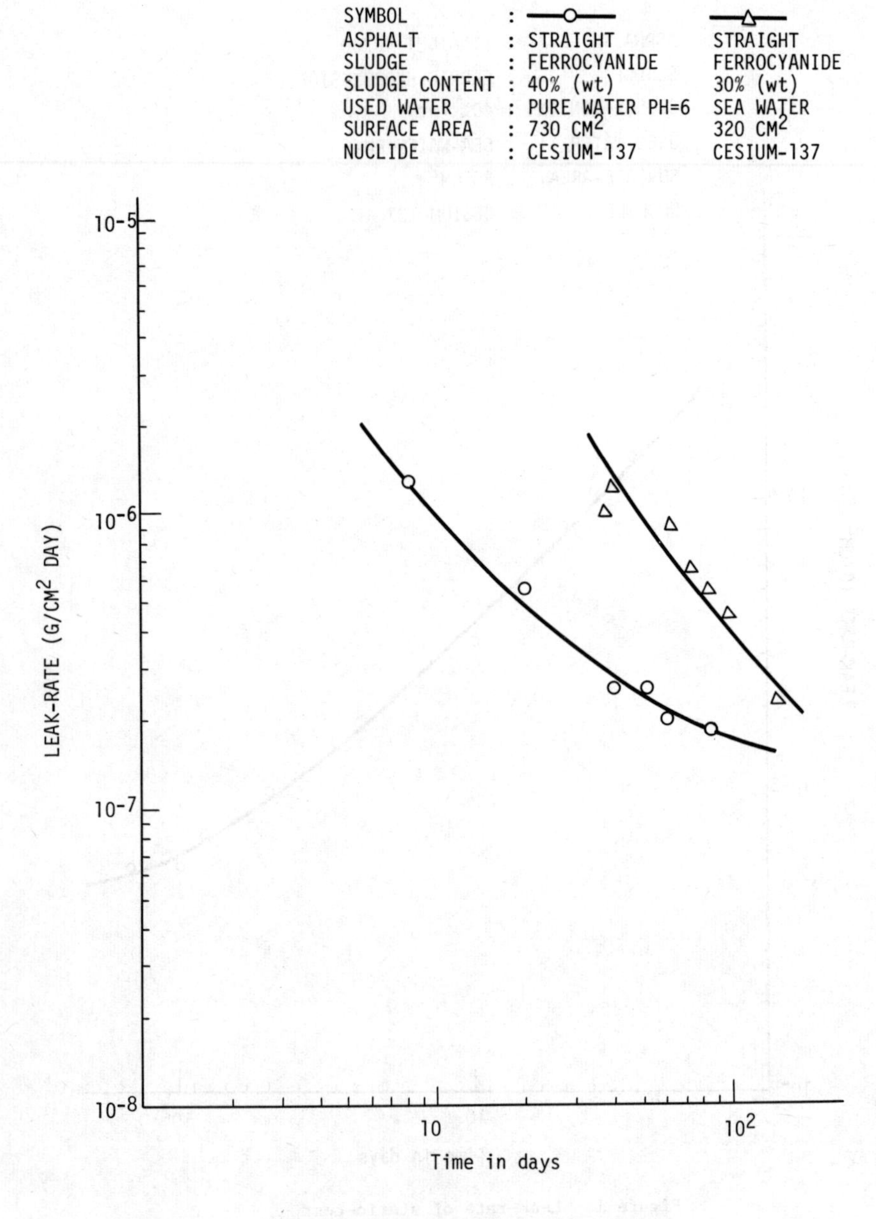

Figure 2 Leak-rate of dynamic-test

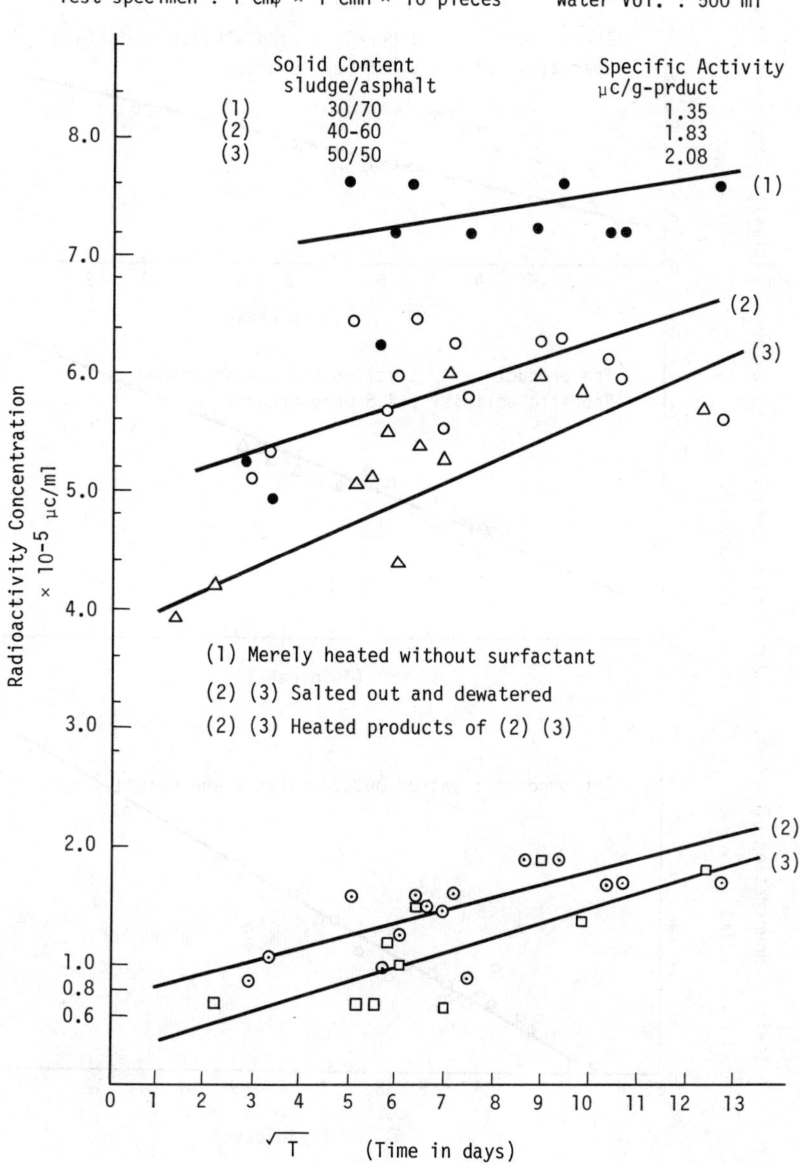

Figure 3 Diffusion of ^{137}Cs of the asphalt sludge mixture into tap water

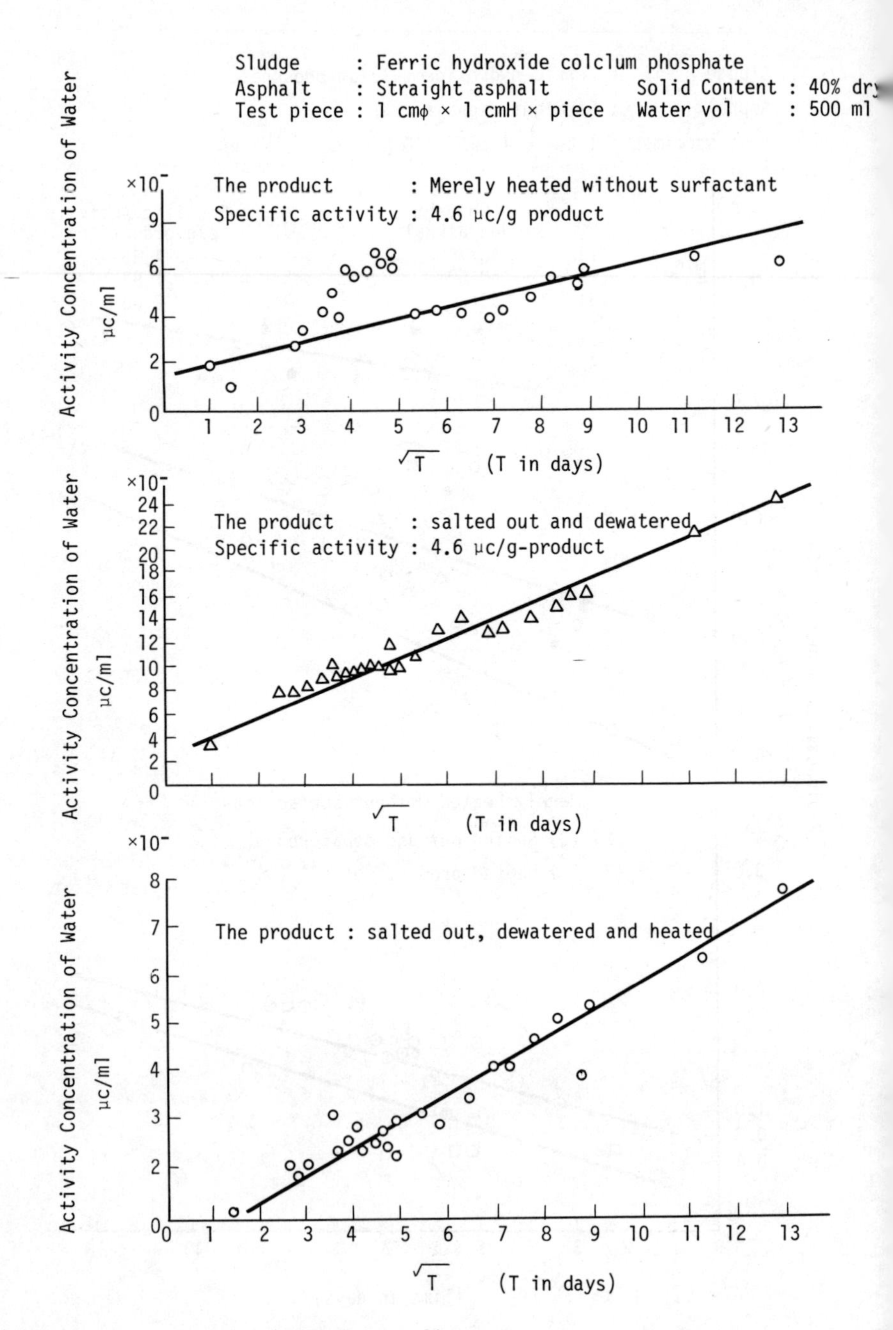

Figure 4 Diffusion of ^{90}Sr of the asphalt sludge mixture into tap water.

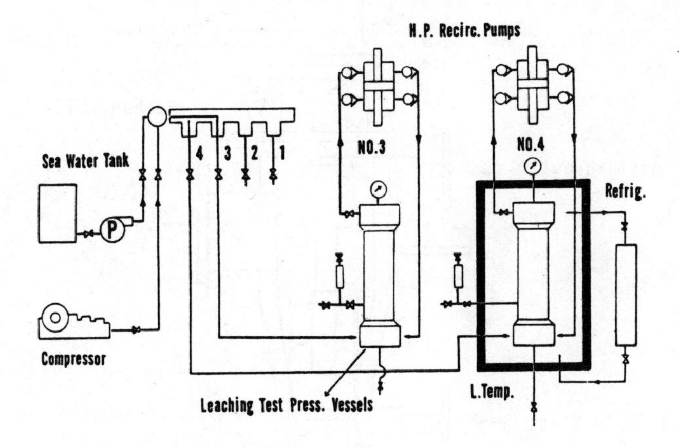

Figure 5. Leaching test flow diagram.

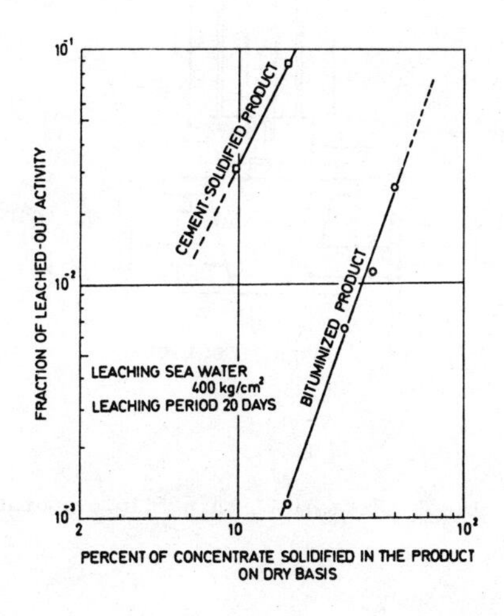

Figure 6 Relationship between fraction of leached-out ^{137}Cs activity and percentage of evaporator-concentrate solidified.

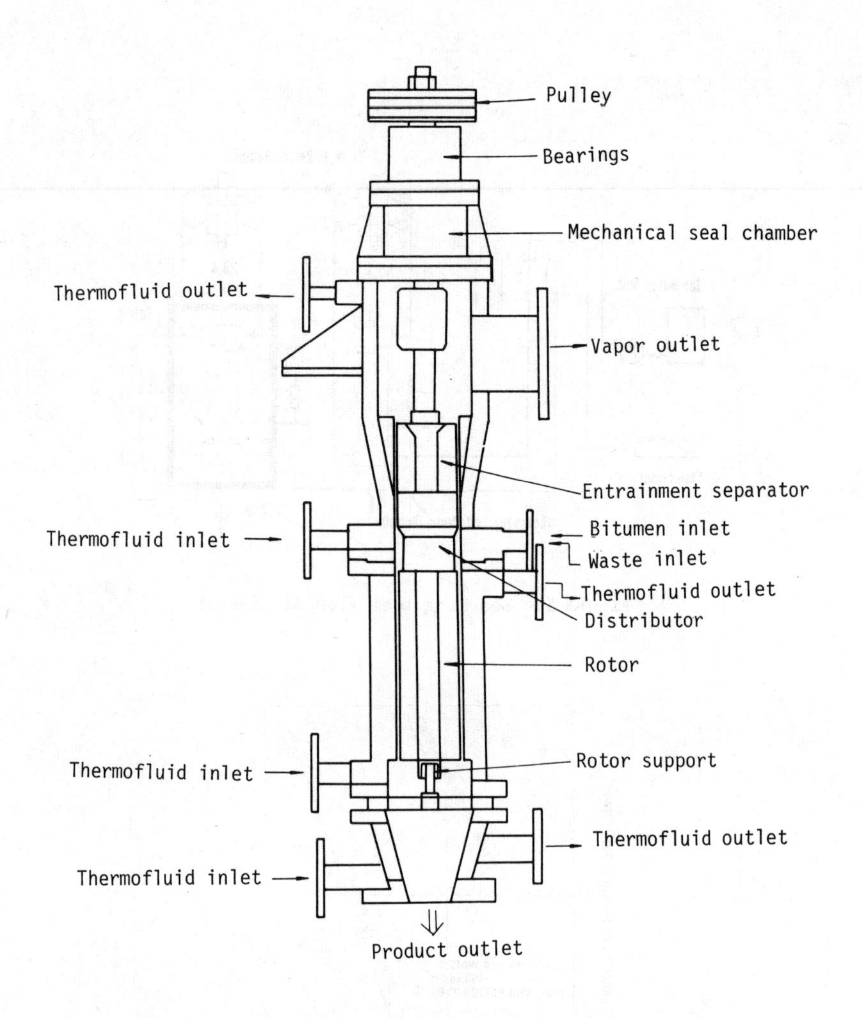

Figure 7 Hitachi-VL thin film evaporator.

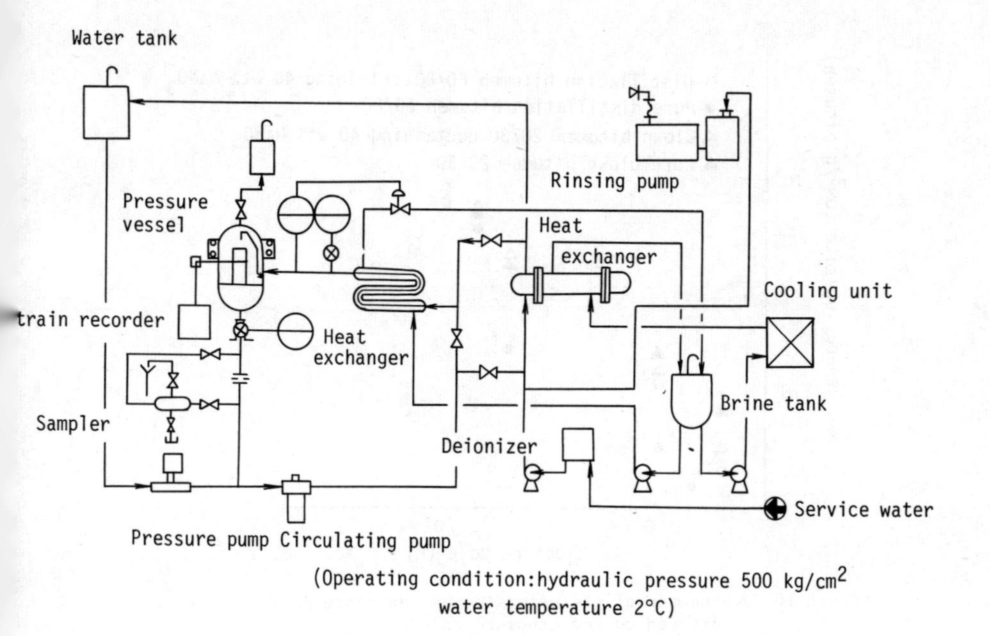

Figure 8 Flowsheet of high pressure leaching test apparatus for real size waste solid

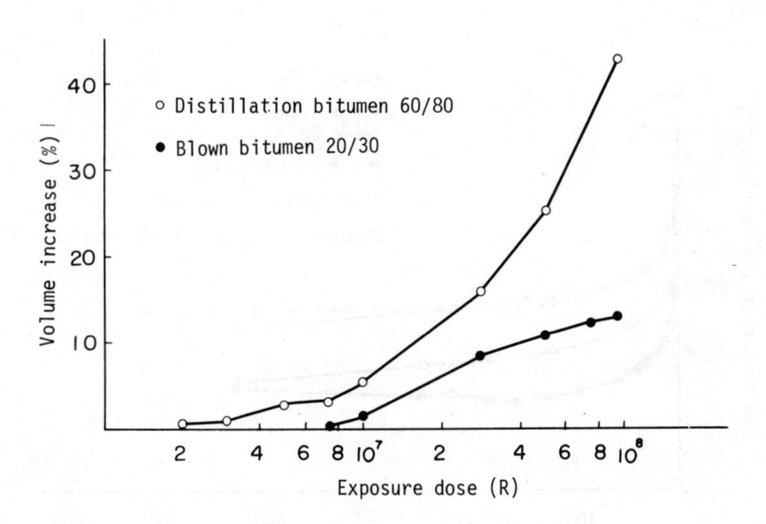

Figure 9 Dependence of volume increase of bitumen-$NaNO_3$ mixture containing 50 wt% sodium nitrate on the exposure dose.

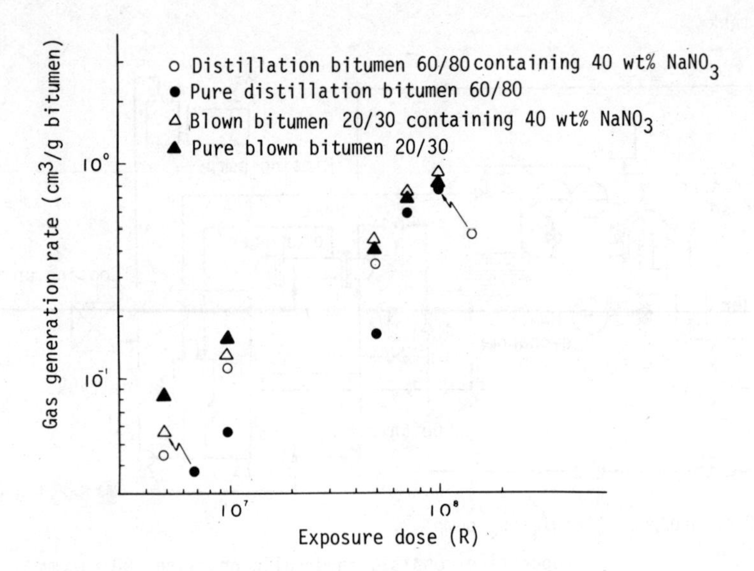

Figure 10 Dependence of gas generated by the radiolysis of bitumen on the exposure dose.

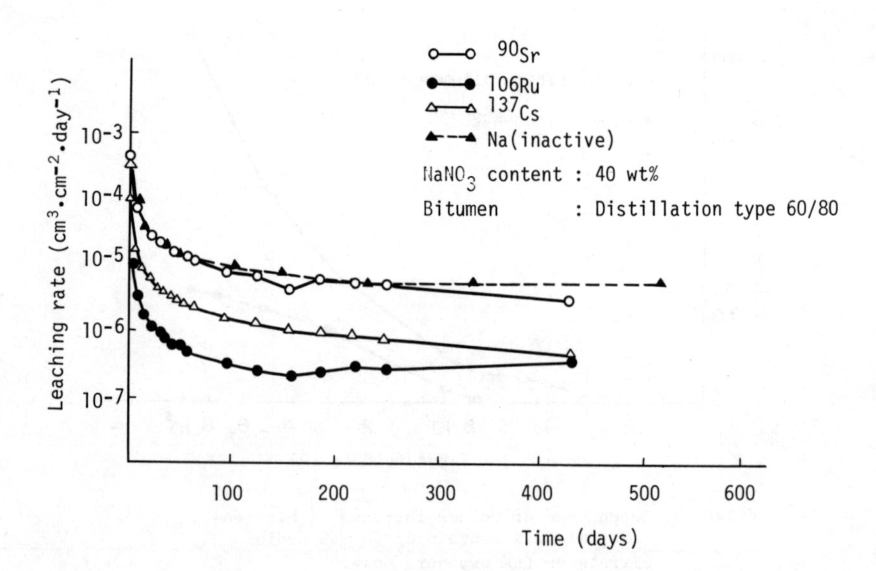

Figure 11. Leaching characteristics of bituminized product.

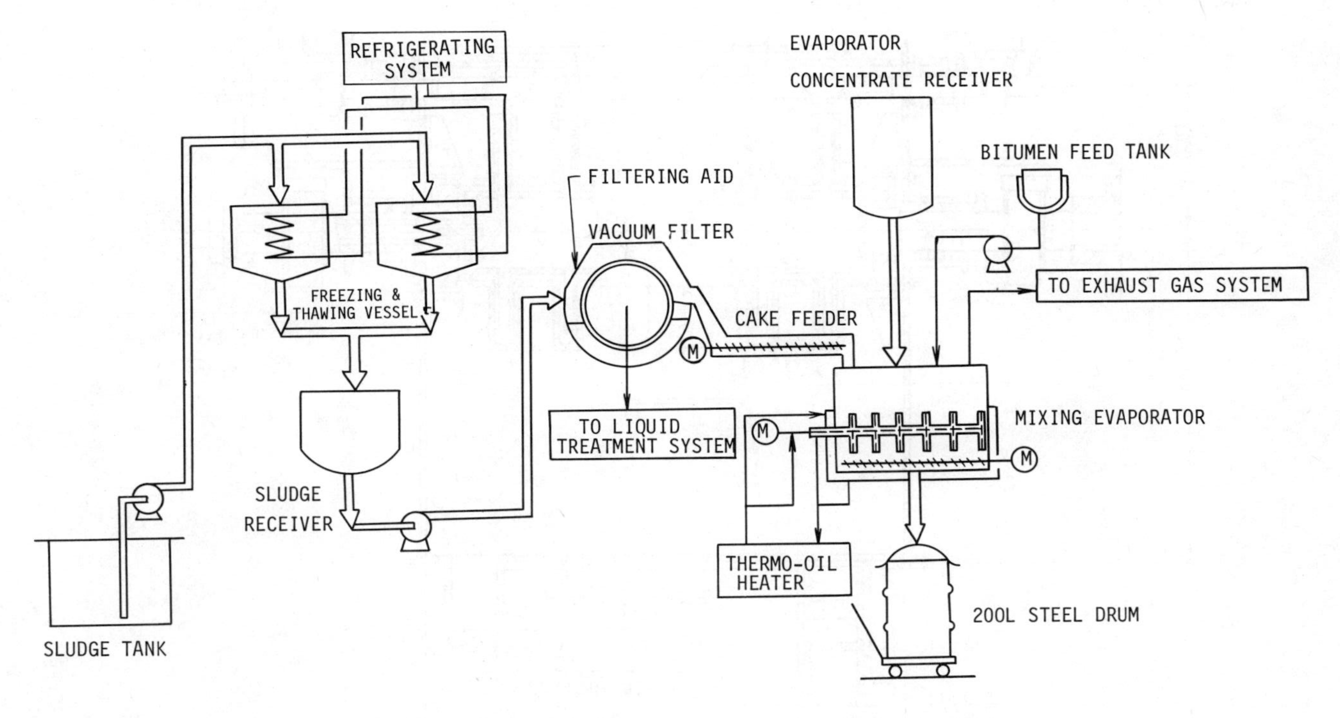

Figure 12 Flow diagram of bituminization equipment

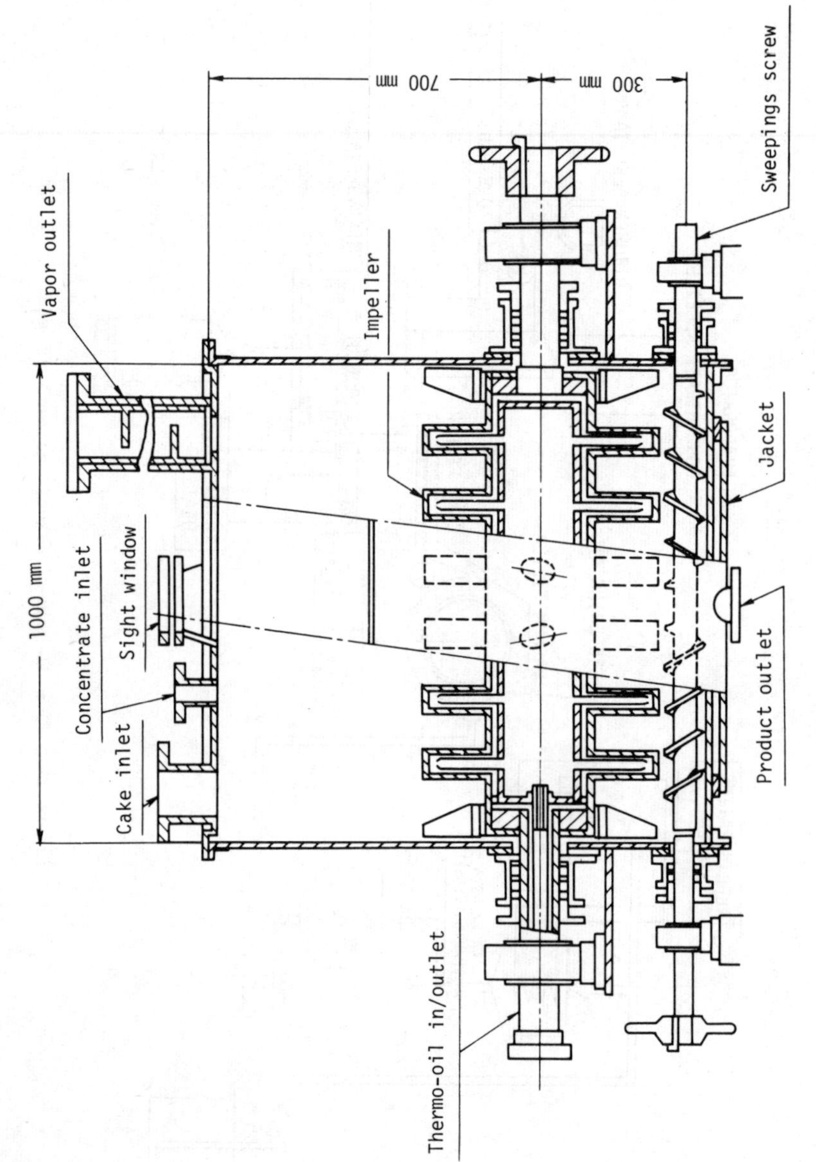

Figure 13 Batch-wise bituminizing unit

REVIEW OF THE RESEARCH AND DEVELOPMENT WORK AND EXPERIENCE IN THE FIELD OF BITUMINIZATION IN THE MEMBER COUNTRIES OF THE COUNCIL FOR MUTUAL ECONOMIC ASSISTANCE

E. Malášek, Czechoslovak Atomic Energy Commission, Prague, ČSSR

and

V. V. Kulichenko, USSR State Committee on the Utilization of Atomic Energy, Moscow, USSR

Review paper describes briefly the research and development work carried out in the member countries of the Council for Mutual Economic Assistance in connection with the development of the bitumization process for the solidification of radioactive wastes. Further on the operational techniques and developed equipment are described and the principles of safe disposal of bituminous products are summarized.

Rapport synoptique qui décrit dans une forme brève les travaux de recherche et développment réalisés dans le cadre des membres du Conseil d'Entr'aide Économique en connexion avec le développment d'un procédé bitumineux pour solidification des déchets radioactifs. Ci-après ce sont décrits les procédés de fabrication ainsi que l'installations technologiques bien developpées et les principes résumés concernant une déposition sûre des blocs bitumineux.

INTRODUCTION

The basic aproach to waste management in socialist countries is to concentrate the activity in the minimum possible volume and to store the wastes in solidified form in waste disposal areas. Therefore great attention was always given to the solidification processes. Bituminization of radioactive wastes has been especially discused in the symposium in 1972 /Development in the Management of Liquid, Solid and Gaseous Radioactive Wastes and Decontamination of Contaminated Surfaces - Warsaw 1973/ and in two panels in 1974 and 1976 /Research of conditions for bituminization of wastes in laboratory scale - Brno and Development of equipment for the bituminization process - Budapest/ [1].

Application of bitumen for solidification of radioactive wastes is based on the following facts:

- the process temperature assures practically complete dewatering and minimum volatility of radionuclides and oil components of bitumen;
- viscosity of bitumen is satisfactory for favourable mixing with the components of wastes and discharging of bituminous product;
- bituminous products containing radionuclides have satisfactory radiation stability and water-resistance for disposal of wastes in simple burial grounds /up to 10^3 Ci/m^3/;
- in contradistinction to cementation the bituminization process does not increase the volume of disposed wastes;
- for the bituminization process cheap and easy available material can be used.

Together with the bituminization process itself great attention has been given to the selection of safe conditions for the burial of bituminous products.

DEVELOPMENT OF THE BITUMINIZATION PROCESS

Development work included the suitability of different types of bitumen, composition of wastes suitable for bituminization /specific activity, chemical composition - maximum content of sodium nitrate, boric acid, manganese, soluble salts, chemical sludges, filtration and ionexchange material/, radiation stability and leaching rate of bituminous products.

Properties of bitumen

Bitumen is a mixture of hydrocarbons, oxidation products of unsaturated hydrocarbons and polymerization products. Composition and properties of bitumen depends on the composition of petroleum and the method of bitumen production. Basic elements of bitumen are 82-88% of carbon, 8-10% of hydrogen, 1-5% of sulphur and 0.5-4% of oxygen.

In Table I. are summarized the types of bitumen, studied in socialist countries.

In Czechoslovakia bitumen is applied in form of emulsion. Two industrial bituminous emulsions were selected [2]:

- cationic emulsion EAS - KS 65 /65% of Romashkin-Saratov bitumen, 0.8% of emulsifier DUOMEEN, 0.4% of HCl and 33.8% of water/,
- anionic emulsion EAS - AS 60 /57% of Romashkin-Saratov bitumen, 3% of emulsifier DEHET D II, 0.5% of NaOH and 39.5% of water/.

Composition of wastes

Bituminization process is applicable to radioactive concentrates of different origin. Therefore the influence of wide scale of salts and their mixtures was studied, including nitrates of

Table I. Types of bitumen suitable to solidification of radioactive wastes

country	bitumen	softening point /°C/	penetration at 25°C /mm/
Bulgaria	"rubrax"	130	1,3
	hydroinsulating bit.	78	2.6
	accumulator bitumen	113	1.6
Czechoslovakia	Romashkin-Saratov bit.	43 - 45	14.3
German dem.rep.	B - 45	54 - 60	3.5 - 5.0
Hungary	NB - 30	70	
	UB - 45	40	
	Romashkin bitumen	58	
Poland	P - 60	60 - 69	1.5 - 4.0
	D - 35	50 - 65	3.0 - 4.0
USSR	BNK - II	40	14.0
	BN - II	40	8.1 - 12.0
	BN - III	45	4.1 - 8.0
	BN - IV	70	2.1 - 4.0

sodium, potassium, aluminium and iron, $CaSO_4$, $CaCO_3$, calcium phosphate, sulphates and chlorides of iron and aluminium, barium sulphate, hydroxides of iron and aluminium, compounds of manganese, boric acid and its compounds, sodium perchlorate, sulphonic acid and inorganic and organic sorbents. The programme covered up reactions between additives and bitumen, thermal and chemical stability, behaviour of bituminous material in contact with water and permissible range of salt concentrations.

For mixtures of different types of bitumen with 20-80% of sodium nitrate the exothermic reaction started at temperatures above 400°C. The presence of alkali /$NaNO_3$:NaOH = 9:1/ decreased the critical temperature to 230-280°C. The presence of emulgators in the mixture of bitumen and 25-70% of sodium nitrate decreased the burning point from 390°C to 270°C. Incorporation of sodium nitrate into bitumen at temperatures not exceeding 200°C is not accompanied with chemical reactions and the bond of bitumen and sodium nitrate has an adsorption character.

Utilization of bitumen with penetration above 14 mm and softening point below 40°C permited the incorporation of 70% of sodium nitrate. Application of harder types of bitumen decreased the incorporated amount of sodium nitrate to 25-30% and it was not possible to realize the process at temperatures below 160°C.

During the bituminization of boric acid at a temperature of 150°C to 175°C the concentration of boric acid in gaseous phase reached 20-25wt%. In the reaction of bitumen with boric acid alkylborate is formed and its oxidation causes the increase of softening point /for BN - II from 41°C to 45°C/ and decrease of penetration /for BN - II from 34 mm to 14 mm/. Addition of borax solution resulted in rapid hardening, maximum content of borax in bitumen reached 20%. Bituminization of borax and sodium nitrate resulted in the formation of bituminous product, containing 30% of bitumen BNK - II, 10% of anhydrous borax and 60% of sodium nitrate. Leaching rate of the product containing 80% of bitumen BN - III and 20% of borax was 3×10^{-3} $gcm^{-2}d^{-1}$. Utilization of bitumen BN - II decreased the leaching rate to 7 x 10^{-4} $gcm^{-2}d^{-1}$ and after 240 days further on to 1 x 10^{-4} $gcm^{-2}d^{-1}$.

Potassium permanganate, used in some decontamination solutions, was a cause of variable manganese salts formation in radioactive waste. Most important of them is manganese nitrate which reacts with bitumen at temperatures below 200°C. Partial reactions of components occured already during the bituminization of manganese sludges at a temperature of 150°C and the compounds created were easily oxidable than the original bitumen. But the yield of such reactions is relatively small and it increases with the extention of contact and increase of temperature. Nevertheless, bituminization of manganese sludges at a temperature of 150°C requires very careful control of the heating system of bituminator, preventing any local overheating above 200°C.

Properties of bituminous material, containing different soluble salts, are given in Table II. Distinct increase of softening point in the presence of aluminium nitrate and ferric nitrate is connected with low decomposition temperature of these salts into oxides /130°C for aluminium nitrate and 75-110°C for ferric nitrate/. The decomposition is connected with an increase of viscosity and formation of gaseous phase containing N_2, NO and partially CO and CO_2. Therefore incorporation of ferric and aluminium nitrates requires the application of bitumen with minimum softening point 50°C and minimum penetration 10 mm or conditioning of concentrates to p_H = 8 before bituminization. Calcium nitrate reacts with bitumen at a temperature of 150-160°C, bituminization at temperatures above 200°C is therefore difficult. Incorporation of calcium nitrate into bitumen at temperatures 150-180°C increases the capture of calcium and strontium.

Table II. Properties of bituminous material, containing 20% of salts /without crystal water/ and 80% of bitumen BNK-II

salt	incorporation temperature			
	150°C		230°C	
	softening point /°C/	penetration /mm/	softening point /°C/	penetration /mm/
$NaNO_3$	48	13.6	57	9.7
$Ca/NO_3/_2.4H_2O$	48	14.7	128	2.0
$Fe/NO_3/_3.9H_2O$	126	0.8	200	clodding
$Al/NO_3/_3.9H_2O$	114	1.1	200	clodding
$FeSO_4.7H_2O$	47	14.2	57	7.1
$Fe_2/SO_4/_3.9H_2O$	66	8.0	138	1.2
$Al_2/SO_4/_3.18H_2O$	55	3.2	98	1.1
$FeCl_3.6H_2O$	58	11.8	141	1.1
$AlCl_3.6H_2O$	84	4.6	93	2.0

Incorporation of sludges into bitumen is not connected with any chemical reactions. Diatomaceous earth from filters was incorporated into bitumen up to 15% with additional 20% of sodium nitrate. Higher content of diatomaceous earth resulted in rapid thickening of the mixture.

Radiation effects

The stability of bitumen with respect to radiation was an important part of studies determining the permissible activity and the burial conditions. Studies of the radiation stability of the bituminous materials were effected by observing of bituminous blocks prepared from real wastes or by irradiation of inactive samples and included namely gas release, volume increase and changes in leaching rate.

Gas release from material containing 60% of bitumen BN-III and 40% of sodium nitrate with various specific activity of ^{90}Sr /namely 0.15 Ci/kg, 1.54 Ci/kg and 15.4 Ci/kg/ has been studied for more than five years. Figure 1. shows the changes of pressure in sealed vessels containing bituminous materials. Material of specific activity 0.15 Ci/kg became radiation stable after two years of storage, but in this case pressure increase caused by gas release was lower than pressure decrease caused by absorption of oxygen from the air. In case of samples 1.54 Ci/kg and 15.4 Ci/kg the gas release continued but the speed of release gradually fell down. This is also a proof of increased radiation stability. Decrease in the speed of gas release was more expressive for sample 15.4 Ci/kg than for sample 1.54 Ci/kg.

Figure 1. Gas release from bituminous materials /60% of bitumen BN-III and 40% of $NaNO_3$/ with various specific activities of ^{90}Sr

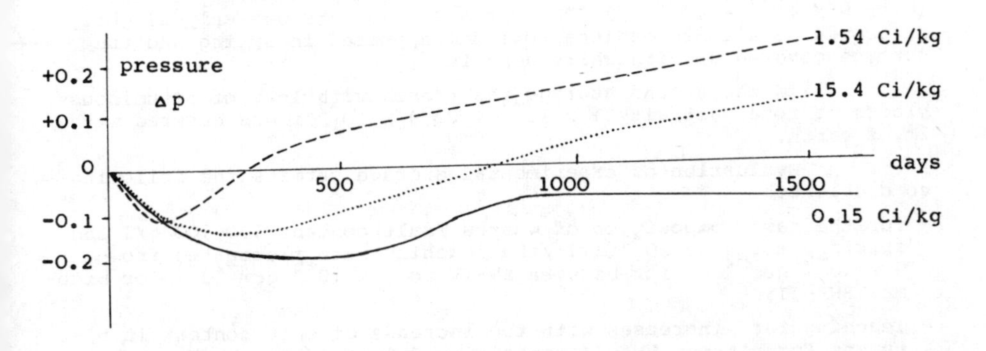

Bitumen BN-III and bituminous material containing 60% of bitumen BN-III and 40% of sodium nitrate were irradiated with ^{60}Co, the dose 2.1×10^5 rad/hour corresponded to about 73 Ci of ^{90}Sr in one kilogram of material. The following composition of gaseous phase was determined /vol%/:

	pure BN-III	60% BN-III+40% $NaNO_3$
H_2	10.6	15.2
CH_4	2.5	2.5
$C_2H_6 + C_2H_4$	0.9	1.0
CO_2	0.13	0.14

Decrease of dose to 2.4 x 10^4 rad/hour /10 Ci/kg ^{90}Sr/ caused the decrease in hydrogen concentration /3.5 vol%/ and gaseous hydrocarbons with CO_2 practically disappeared.

In other experiments the changes of bitumen constituents caused by irradiation were studied. These irradiation experiments lead to the following conclusions:

- in case of bituminous materials of specific activity below 1 Ci/kg radiation damage is very small and the release of gaseous radiolytic products is negligible;
- materials of specific activity up to 10 Ci/kg releases small amount of hydrogen, but for bitumen BN-III the concentration of hydrogen in gaseous radiolytic products is below explosive concentration / < 4 vol%/;
- with an increase of specific activity above 10 Ci/kg the amount of explosive gases /H_2, CH_4, C_2H_4, C_2H_6/ becomes greater and therefore in selection of conditions for safe disposal the removal of radiolytic gases should be considered.

Leaching of bituminous products

The main results of extensive laboratory experiments with a view to leach rate of bituminous products have been already published [3,4]. Recent programme includes the long term storage experiments in open air and in the ground.

Bituminous blocks /30 cm x 30 cm x 8 cm/ were produced from bitumens BNK-II, BN-II, BN-III and BN-IV and contained 27-60% of salts.

Characteristic feature of the storage in open air is an oscillation of activity during the year. Maximum specific activity is leached in autumn and in spring. Depending on the specific activity of bituminous blocks the maximum total β-activity in the leach was 10^{-7} - 10^{-9} Ci/l. The surface of blocks was only slightly destroyed. Small depressions /0.1-3mm/appeared in spring and the surface covered up with white deposit.

In the ground storage the trench with 1-2t of bituminous blocks of total activity 1 x 10^{-2} Ci to 9 x 10^{-2} Ci was covered with 2m of earth.

Evaluation of experimental storage permits the following conclusions:

- for the same composition of wastes /salt content 500-700 g/l and specific activity 10^{-5} Ci/l /the leaching rate fluctuated from 2 x 10^{-4} $gcm^{-2}d^{-1}$ for bitumen BN-IV to 6 x 10^{-5} $gcm^{-2}d^{-1}$ for bitumen BNK-II;
- leaching rate increases with the increase of salt content in bitumen. For bitumen BNK-II containing 35% of salts the leaching rate was 1.5 x 10^{-4} $gcm^{-2}d^{-1}$, for 45.3% of salts leaching rate of 1.7 x 10^{-4} $gcm^{-2}d^{-1}$ and for 53% of salts leaching rate of 2.4 x 10^{-4} $gcm^{-2}d^{-1}$;
- when soft bitumen /BN-III, BNK-II/ is used for incorporation of wastes containing sodium nitrate, the optimum content is 40% of $NaNO_3$;
- presence of alkali in the amount exceeding 5% decreases the water resistance;
- presence of nitric acid in the amount up to 10 g/l in bituminous material containing sodium nitrate does not effect the water resistence;
- leaching of ^{137}Cs and ^{90}Sr depends on the amount of incorporated wastes /sodium nitrate/, solubility of compounds containing

radionuclides and on sorption of radionuclides on the particles of filling;

- leaching rates of bituminous material established in laboratory scale are of the same order as in an experimental storage;
- the highest leaching rates were observed for bitumen BN-IV and the lowest ones for soft bitumens BN-II and BN-III;
- the depth of penetration of water into bituminous blocks in case of storage in water or in wet ground is not directly proportional to time.

OPERATIONAL TECHNIQUES

The technological processes for the bituminization of radioactive wastes are similar to the processes developed in western Europe. The main differences are a consequences of the utilization of locally produced material and equipment and variations in the amount and composition of wastes.

Very simple bituminization of natural sorbents was developed in Bulgaria. Natural sorbents are filled into bitumen moulds, working as a system of four sorption columns. The unit has a capacity of 5 l/h and it is used for decontamination of water containing less than 1 g/l of salts and 10^{-4} - 10^{-9} Ci/l of different radionuclides. Selection of natural sorbents depends on the composition of wastes and the decontamination factor reaches 3 x 10^{3} to 10^{4}. When saturated natural sorbents are buried bitumen ensures water impermeability.

In Czechoslovakia the first bituminization unit was a pilot plant with the capacity of 40 l/week of bituminous product containing an activity up to 20 Ci/kg. The unit is still used for discontinuous solidification of small amounts of medium active wastes. Radioactive concentrates and bitumen emulsion in water are fed into mixed kettle evaporator. The kettle evaporator has duplicated walls with oil heating. The mixture is mixed intensively and heated up to 140°C. The foam is completely destroyed with a help of cooling circuit located above the liquid level in evaporator. After evaporation, the mixture is discharged into cast forms from stainless steel, cooled and transported to waste disposal area.

For the treatment of low active concentrates and sludges at NRI Řež a continuous bituminization unit has been developed. Radioactive concentrates and bitumen emulsion in water are treated in a wiped film evaporator at a temperature of 150°C. The vapors and gases pass through an aerosol separator to condenser. Condensated water is discharged after monitoring and remaining gases pass through an absolute filter and stack to the atmosfere. Bituminous product drops from the bottom of evaporator into a steel or paperboard storage container. The unit is used for bituminization of concentrates containing 500 g/l of salts, the bituminous product contains about 40 wt% of salts. The capacity is about 50 l/h of evaporated water.

For the treatment of evaporator concentrates and saturated ionexchange materials the bituminization unit with capacity of 200 l/h of evaporated water is under development, a verified prototype will be available in 1978. The technology is based on the application of a bituminous emulsion and a wiped film evaporator.

In the German democratic republic a discontinuous pilot plant bituminization unit has been constructed and examined. The unit is composed of a heated vessel for storage of melted bitumen at a temperature of 150°C, a storage vessel for radioactive concentrates, mixed kettle bituminator and a system of gas purification. The unit has a capacity of 50 l/h of evaporated water at a working temperature 200-230°C, the bituminous product contains up to 50% of salts. The pilot plant is used for bituminization of evaporator concentrates, ion exchange materials and activated charcoal.

In Hungary a pilot plant bituminization unit is based on heating of bitumen in a relatively narrow reactor to a temperature of 250°C. The bitumen is mixed with nitrogen and introduced at the bottom of the bituminator. The concentrate is poured drop by drop at a slow rate into the surface of the hot bitumen. On contact with the bitumen the concentrate loses its water content and the salt is mixed with the bitumen with a very fine distribution. The capacity is up to 10 l/h of evaporated water and the bituminous product contains 25% of salts. The unit can be used in small installations producing 10-15 m^3/year of concentrates.

The bituminization plant in Poland is based on Belgian experience and operates with a non-emulsified bitumen with an evaporating capacity up to 50 l/h. Radioactive concentrates are fed into a mixer-evaporator. The evaporation of water and the mixing of the solid residue with molten bitumen takes place at a temperature between 200°C and 230°C and is accompanied by vigorous stirring. The bituminator is electricaly heated, the heating capacity is satisfactory to maintain the temperature inside the mixture up to 340°C. The process is discontinuous. It is used for solidification of radioactive concentrates with an activity up to 10^{-4} Ci/l at the nuclear research institute. It has been also verified for bituminization of chemical sludges, barium carbonate, cupric sulphate and potassium hexacyanoferrate.

The most extensive volume of development work has been conducted in the USSR.

The first bituminization unit has a capacity of 70 l/h of evaporated water. The bituminization process is conducted in a reactor - mixer, into which the molten bitumen is fed from the melter along with a measured amount of evaporator concentrate. As soon as the required salt content has been attained, the bitumen mass is discharged into a storage container. The evaporated water pass through a separator to condenser. The unit is suitable for smaller sources of concentrates /up to 50 l/h/.

Different principles have been applied for continuous bituminization process with the capacity of 200 l/h of evaporated water. In the first step the water is removed from the concentrates in a dewatering unit at a temperature of 130°C. Dry product is fed into an extruder together with the bitumen heated to 160°C. The bituminous product contains up to 70% of salts and it is discharged from the extruder into a storage container at a temperature of 130°C. The unit is suitable for solidification of concentrates at a nuclear power plant with PWR reactor.

Prevailing part of installations is based on wiped film evaporators. The whole set of units is available with capacities 50 l/h, 150 l/h, 230 l/h and 500 l/h of evaporated water. Standard equipment includes storage vessels of radioactive wastes and bitumen, dosing pumps for bitumen and concentrates, condensers, wiped film evaporators etc. The film evaporator consists of a heated cylindrical vertical body. The concentrates and melted bitumen are separately dosed and transported by dosing pumps into the upper part of the evaporator through a distribution ring. The products are seized by the blades of the rotor, distributed and uniformly laid on the heated surface in the form of a thin film. The water is removed from the mixture in the lower part of the evaporator at a working temperature of 160°C. Water vapour flows up, passes through the separator and is afterwards condensed. The final product contains 50-60% of salts. The process is continuous, but it can be operated also discontinuously. The system is suitable for solidification of concentrates at nuclear power plants.

Two steps solidification unit is also used at the Central waste treatment plant in Moscow. Radioactive concentrates are fed into rotary dryier heated up to 650°C. Gaseous phase is purified in a system of cyclone separators, a condenser and absolute filters. Dry product is fed into an extruder together with the bitumen heated to 60°C. The bituminous product contains up to 80% of salts and it is discharged into a transportation container of the volume 0.5 m^3. The process is suitable for solidification of low active concentrates without the requirements of shielding.

SAFE DISPOSAL OF BITUMINOUS PRODUCTS

The basic factors which influence the safe disposal of bitumen blocks containing radioactive wastes are the sort of bitumen, chemical composition of wastes and thermal and radiation stability of blocks. The most important factor is the leaching rate of the radionuclides in water. Disposal of solidified wastes in water or in wet ground is viewed upon as a version of disposal without hydroinsulation, for no shell can be expected to withstand the effects of corrosion for any appreciable length of time.

In the member countries of the Council for Mutual Economic Assistance "Methodology for selection of safety conditions for burial of solidified wastes in relation to their properties and specific activity" has been worked out and approved.

Selection of a burial ground requires:

- the determination of the necessity to hydroinsulate the burial place;
- the determination of the necessity to seal the burial place with the aim to prevent the extention of radionuclides from the storage place out of the contact with water;
- determination of the possibility to seal solidified wastes;
- determination of maximum permissible temperature of storage;
- the determination of the necessity to remove the heat and selection of conditions for heat removal.

In case of bituminous material the heat release, possible temperature of storage and heat removal are not taken into account. Standard methods of preparation and examination of samples and mathematical methods for evaluation of results are used in these studies. In hydrogeological studies, dilution of seepage water in ground water and sorption of radionuclides on the rock mineral are not taken into account.

Hydroinsulation is not necessary in case of

- burial places located in dry places with natural hydroinsulation /salt mines/ or in places, where the average amount of rainfall water is lower than an average evaporation rate;
- burial places located above the ground water level, where the seepage water in time of reaching the ground water contains radionuclides in concentrations below the MPC levels.

REFERENCES

/1/ Tolpygo, V. and Fridrich, B.: "On cooperation of member countries of the Council for Mutual Economic Assistance in the field of reprocessing and burial of radioactive wastes", Proc. Internat. Symp. Management of Radioactive Wastes from the Nuclear Fuel Cycle, IAEA, Vienna 1976

/2/ Malášek, E. and Nápravník, J.: "Improvement in the solidification processes and the influence on the burial conditions", IAEA Panel on the Choice of Burial Conditions for Radioactive Concentrates as a Function of their Properties and Activity Level, Moscow, 1972

/3/ Bituminization of Radioactive Wastes, Technical Reports Series No. 116, IAEA, Vienna, 1970

/4/ Waste Management Techniques and Programmes in Czechoslovakia, Poland and in the Soviet Union, IAEA, Vienna, 1971

Discussion

E.P. UERPMANN, F.R. of Germany

M. Malášek, do you have an explanation for the decrease and following increase of the curve shown in Figure 1 ? Why is the curve with the highest activity per unit volume lower than that with the activity ten times less ?

E. MALASEK, Czechoslovakia

The changes of pressure are influenced by two main reactions - absorption of oxygen in bitumen and release of gaseous radiolytic products. At the beginning of storage, the first reaction is prevailing and causes the decrease of pressure, which is gradually compensated with the increasing volume or radiolytic products.

In the case of the highest activity, a slight increase in the volume of bitumen was observed and therefore all radiolytic gases were not released.

W. HILD, F.R. of Germany

I would like to say that in our external irradiation experiments, we have also noticed a considerable O_2-depletion. Could you kindly tell us whether you can derive from your irradiation tests a specific radiolytic gas production rate, like for instance volume of radiolytic gas produced per gram of product and 100 Mrad absorbed dose ?

E. MALASEK, Czechoslovakia

The radiolytic gas production depends on many factors, for example on the type of bitumen, the composition of the bituminous products, the type of radiation, etc. Therefore, it is not possible to give any general figure.

G. LEFILLATRE, France

What do you think about the difference in the results obtained from auto-irradiation experiments made in the US(ORNL) and the USSR (in the case of the US experiments with 13.9 Ci/l of fission products there was no release of gas from radiolysis ; in the case of the USSR, with ^{90}Sr at 354 Ci/kg and 15.4 Ci/kg there was a release of Ar and other gases from radiolysis.

E. MALASEK, Czechoslovakia

It rather difficult to comment on this. I can only say that in the case of experiments in the Soviet Union it was a pure bitumen containing only sodium nitrate in the amount of 40 % and ^{90}Sr. I do not know the exact composition of the wastes which were used in the United States but the reason might be the different radiochemical reactions of the components which were present in both cases.

Y. SOUSSELIER, France

In your paper you have mentioned studies on a fairly large number of bitumen qualities. I should like to know :

1) If this bitumen came from crude oils with very different chemical characteristics or not.

2) If you think that the results justify that systematic studies should be carried out on the various possible types of bitumen obtained from very different crude oils.

E. MALASEK, Czechoslovakia

The main difference in the chemical composition of different types of bitumen is in the content of sulphur. The increased content of sulphur is typical for Romashkin-Saratov bitumen. The properties are mainly influenced by the methods of preparation. For example, BN-III and BNK-II are soft bitumens, BN-IV is a hard type of bitumen.

In the Member countries of the Council for Mutual Economic Assistance, the studies of the properties of bitumen and bituminous products were practically terminated several years ago and at present only development of equipment is carried on.

V. KLÜGER, F.R. of Germany

We have also made irradiation experiments and we think that the volume increase of the samples is not a good method to decide whether the products are stable or not. We have made irradiation experiments with products from German reactor waste which were irradiated to a total of 80 Megarads and we found a porosity of only 1 to 3 %.

Discussion

E.P. UERPMANN, R.F. d'Allemagne

M. Malášek, pouvez-vous expliquer l'abaissement et le redressement consécutif de la courbe qui apparaît à la Figure 1 ? Pourquoi la courbe correspondant à la plus forte activité par unité de volume est-elle plus basse que celle correspondant à une activité de dix fois inférieure ?

E. MALASEK, Tchécoslovaquie

Les variations de pression sont influencées par deux réactions principales, soit l'absorption d'oxygène dans le bitume et la libération de produits gazeux de radiolyse. Au début du stockag la première réaction l'emporte et entraîne une diminution de la pression, qui est progressivement compensée par le volume croissant de produits de radiolyse.

Dans le cas de l'activité maximale, on a observé une faible augmentation du volume de bitume, aussi tous les gaz de radiolyse n'ont-ils pas été libérés.

W. HILD, R.F. d'Allemagne

Je souhaiterais signaler qu'au cours de nos expériences d'irradiation externe nous avons aussi observé un appauvrissement très important en O_2. Auriez-vous l'obligeance de nous dire si vous pouvez déduire de vos essais d'irradiation un taux de production spécifique des gaz de radiolyse, tel que, notamment, le volume des gaz de radiolyse obtenu pour un gramme de produit et 100 Mrads de dose absorbée ?

E. MALASEK, Tchécoslovaquie

La production de gaz de radiolyse dépend de nombreux facteurs, et notamment du type de bitume, de la composition des produits bitumineux et du type de rayonnement. En conséquence, il n'est pas possible de donner un chiffre général.

G. LEFILLATRE, France

Que pensez-vous des différences de résultats observés entre les expériences d'auto-irradiation américaines (ORNL) et russes ? (Dans le cas des américains, avec des produits de fission de 13,9 Ci/l, aucun dégagement de gaz de radiolyse. Dans le cas russe, avec du ^{90}Sr à 354 Ci/kg et 15,4 Ci/kg, dégagement d'Ar et autres gaz de radiolyse).

E. MALASEK, Tchécoslovaquie

Il est assez difficile de formuler des commentaires à ce sujet. Tout ce que je peux dire, c'est que, dans le cas des expériences effectuées en Union Soviétique, il s'agissait de bitume pur ne contenant que du nitrate de sodium dans une proportion de 40 % et du ^{90}Sr.. Je ne connais pas la composition exacte des déchets qui ont été utilisés aux Etats-Unis, mais l'écart pourrait s'expliquer par les différentes réactions radiochimiques des composants présents dans l'un et l'autre cas.

Y. SOUSSELIER, France

Dans votre communication, vous mentionnez des études sur un assez grand nombre de qualités de bitume. Je voudrais savoir :

1) si ces bitumes provenaient de pétrole brut de caractéristiques chimiques très différentes ou non,

2) si vous estimez que les résultats justifient que l'on mène des études systématiques sur les différents types possibles de bitume obtenus à partir de pétrole brut très différents.

E. MALASEK, Tchécoslovaquie

La principale différence dans la composition chimique des divers types de bitume réside dans la teneur en soufre. Le bitume Romashkin-Saratov se caractérise par une plus forte teneur en soufre. Les propriétés sont principalement influencées par les méthodes de préparation. C'est ainsi que le BN-III et le BNK-II sont des bitumes mous, alors que le BN-IV est un bitume de type dur.

Dans les pays membres du COMECOM, les études sur les propriétés du bitume et des produits bitumineux ont été pratiquement achevées il y a plusieurs années et, à l'heure actuelle, on ne procède qu'à des mises au point du matériel.

V. KLÜGER, R.F. d'Allemagne

Nous avons aussi procédé à des expériences d'irradiation et nous estimons que l'augmentation de volume des échantillons n'est pas une bonne méthode pour décider si les produits sont ou non stables. Nous avons effectué des expériences d'irradiation sur des produits tirés de déchets provenant de réacteurs allemands, qui ont été irradiés à une dose s'élevant au total à 80 mégarads, et observé une porosité de 1 à 3 % seulement.

Session III
Panel

Chairman - Président

Y. SOUSSELIER

Séance III
Table ronde

MEMBERS OF THE PANEL

MEMBRES DE LA TABLE RONDE

Mr. E. ESCHRICH, Eurochemic

Mr. W. HILD, F.R. of Germany

Mr. G. LEFILLATRE, France

Dr. E. MALASEK, Czechoslovakia

Mr. Y. SOUSSELIER, France

Mr. N. VAN DE VOORDE, Belgium

Panel

Y. SOUSSELIER, France

Until the opening of this Seminar on the bituminization of low and medium level radioactive wastes, the real value of holding such a meeting might have been queried as it was being held less than one month following the major joint IAEA/NEA Symposium in Vienna on the overall problem of waste, at which the question of bitumen had already been discussed. It could also have been presumed that such a Seminar, to be attended by experts only, might result in some sort of mutual congratulatory exercise or one of general complacency. It is obvious that nothing of the kind occurred. We have all learnt and gained a great deal from our discussions on the papers submitted to us, which were most interesting,and from our comprehensive visit to the Eurochemic facilities. The first lesson, which some of us had already drawn following the Vienna Symposium, is that a meeting is more profitable if it is attended by a small number of experts dealing with familiar subjects and hence able to discuss them in depth.

It is already possible to draw several conclusions or rather, to confirm certain points. The bituminization of low and medium level radioactive waste is a well-established process. Several plants are in operation throughout the world. Results so far are most satisfactory. On the whole there have been very few hitches and the causes of the most important ones (which did not have any major consequences) and the ways in which they can be prevented in future are very well known. Experience gained in the storage of the products obtained is also satisfactory.

Does this mean that our work is now over and all that remains to be done is to build further plants and to run them for the good of nuclear energy ? This is undoubtedly so, but all the same our discussions have revealed that several problems remain unsolved. One example is the fluidity of bitumen and its possible adverse implications for a type of storage such as that currently used by Eurochemic, assuming that after several decades these containers will have to be retrieved and that the drums will have been corroded to some extent or even destroyed by corrosion. Another question is the fire risk. We know that this has been discussed at length in certain countries and it is even likely that some of these, such as the United States, did not start to study bituminization any sooner because they regarded this risk as quite considerable. During our visit to Eurochemic, I heard questions being asked on this maximum risk for storage, for instance in the event of an aircraft crashing on to the storage facilities. Another question asked was whether there was any advantage in using a highly processed product such as bitumen when final storage methods already offering extensive safeguards were available. Other questions concerned the selection of the bitumen grade, and during this Seminar perhaps insufficient emphasis was laid on the fact that the term "bitumen" covers a very extensive category of products with widely differing physical, chemical or physicochemical characteristics, possibly because everyone here was aware of it.

All these questions are interrelated of course, as shown by the discussions. Generally, it must be borne in mind that the various stages in waste management, from production to final storage, interact. If a process is to be optimized, or even selected, it is not enough to examine what is happening at a well-defined point and to effect savings in a restricted field. On the contrary, the nature of the economic assessment as a whole and the cost/profit analysis as a whole for the waste management from the production stage to final storage should be examined within a much more general context. Naturally, this is a difficult task, often even impossible at the moment, because in many of our countries the final storage method has not yet been definitively adopted. Actually, it seems to me that our German colleagues are the only ones who have clearly defined the final storage method which they intend to adopt. In most other countries there are still several options available, several possible solutions.

Nevertheless, without wishing to discuss all these problems of waste management interactions and procedures as a whole, I feel we could deal with several of the points just mentioned. I suggest that we could group them into five categories (this is inevitably a little arbitrary) :

- The first might for instance concern problems connected with bitumen fluidity. Should these be regarded as important ? In the affirmative, is it possible to envisage methods of varying complexity for mitigating the problem ? Must these fluidity problems entail storage decisions such as whether to send waste to final storage as quickly as possible ? Or is there a need for more complex containers than the ones currently used ?

- The second question, which is also very general, concerns product quality. We have heard a great deal about leaching rates. We have seen that they are very good and that in fact they are often better than those originally anticipated The question may still arise, however, as to whether any attempt should be made to improve this property or, on the contrary, once a given value has been reached and in view of the radionuclides present in waste, whether there is any need to continue endeavouring to improve the leaching rate.

- Thirdly, the maximum risk linked to a fire in the course of storage or during transport. Should this risk be taken into consideration and if so, how can it be mitigated, what proposals can be made to reduce it or even remove it ?

- Fourthly, the difficult problem of the overall economics and optimization of the process.

- Finally, the selection of the bitumen and whether there is any point in trying to determine whether or not this choice has much influence on the process, in view of the range of bitumens of all origins and compositions.

I shall first ask the Members of this Panel for their views on these questions then I shall ask Members of the meeting whether they have any questions or comments. I hope that during this Round Table meeting unduly specific questions will be avoided and that the discussions will be relatively general in the light of their possible implications for our R & D programmes.

Let us first deal with fluidity. Is this a major problem ? Can it be avoided ? Should it be taken into account in storage ? As M. Lefillatre asked the first question on storage at Eurochemic, I should like to ask him for his views on the subject.

G. LEFILLATRE, France

I asked the question on interim storage at Eurochemic yesterday because, originally, this storage had been planned for a period of ten years and accordingly, the problem of corrosion of the drums did not arise. This storage time was subsequently increased so that it could possible extend to fifty years. I therefore put this question with the thought in mind that ordinary mild steel drums might be corroded by the end of such a period. I received the reply that as a superior material, chrome steel, had been selected, there was no danger of this happening. Personally, I believe that this problem is encountered at interim storage level, but no longer at final storage level, and this point can be discussed in greater detail later. I fail to see, however, how bitumen fluidity could be prevented, even by adding fillers, which would cause technological difficulties. Since bitumen is plastic, it is bound to flow in the course of time, and the risks of this occurrence must be studied whether final storage is carried out above ground, underground or on the sea-bed. Finally, in view of the product's leaching properties, so long as temperature and self-heating are not a problem (this possibility should be taken into account for the final storage of waste exceeding 1 to 10 Ci/l, especially 10 Ci/l) I think that all my colleagues and myself have made the most pessimistic calculations on the assumption that the containers of single bitumen blocks would be destroyed. In short, I do not think that fluidity risks constitute a problem in final storage.

W. HILD, F.R. of Germany

In my opinion, M. Lefillatre gave us an excellent summary of the situation in this field. Perhaps it could be added that conditions are somewhat favourable because a material with visco-elastic properties is available and by adding solid salts or other products, it is possible to alter the viscosity to a certain extent in order to obtain more stable products with regard to storage at normal temperatures and with regard to fluidity during interim storage. Fluidity cannot be entirely dismissed however because a material remains visco-elastic even if its viscosity has been increased and it must be assumed that during final storage the situation described by M. Lefillatre will be reached. You were able to note from the calculations submitted by Mr Smailos this morning that the possibility of fluidity and total filling of the cavity by the product are being taken into account.

H. ESCHRICH, Eurochemic

The question of how to store final product drums, either in a vertical or in a horizontal position, has naturally a connection with the fluidity of the bitumen product. As far as interim storage is concerned, the fluidity has of course to be taken into account, whether one stores the drums in a vertical or a horizontal position. The duration of interim storage to be foreseen depends on the waste management policy established within the neighbourhood, the country or even the continent ; however, at Eurochemic, for example the interim on-site storage period could not be clearly defined at present. Therefore we had to choose a durable material for our drums (and I come here to the fourth of our five questions which are all interconnected with the economical and technical optimization of the bituminization of wastes) ; for cost reasons we hesitated to choose one of the best materials known to be corrosion-resistant towards the usual environmental media, such as water and air, which is stainless steel. Stainless steel would certainly be a good choice but it is expensive. Mild steel drums protected by a layer of paint are most certainly the cheapest metal containers.

We also considered using galvanised mild steel drums which might - under certain storage conditions - be more corrosion-resistant than painted mild steel drums.

The material we have finally chosen on the basis of various corrosion tests is "chromized steel" ; this is a "sandwich type" material which consists of a thin stainless steel-type layer on both sides and in between a thicker layer of normal mild steel. Its properties and price compared to other candidate materials were reported yesterday.

The drums made of chromized steel appeared to us to be the best compromise between price and required properties. Even if the mild steel component was heavily corroded after a prolonged storage time it could be expected that the drum remains dimensionally stable and that the inner chromized steel layer would remain intact preventing the bitumen product flowing out of its container and permitting the retrieval of the waste by the remotely operated crane or a lift truck without coming into direct contact with its active contents.

In connection with the question of fluidity I would like to recall that the drums are closed by a cover of thirty centimetres diameter, but not gas-tight, to allow the radiolysis gases to escape; if the drums were stored in a horizontal position the risk of clogging the slits by the creeping bitumen would exist and consequently a pressure build-up within the drum and an outflow of bitumen product could occur.

To conclude : if interim storage of an unknown duration has to be envisaged mechanically stable and corrosion-resistant containers should be used and preferably stored in a vertical position

N. VAN DE VOORDE, Belgium

We are discussing the fluidity of bitumen but a number of countries are using pure bitumen to incorporate metal and other solid waste. At this moment the real question is : is this a good method of operation, because this bitumen is really fluid and remains fluid and can be squeezed out like a tube of toothpaste ! Is pure bitumen an acceptable means of incorporation of radioactive heterogenic material ?

Y. SOUSSELIER, France

Am I right in thinking that you are referring to metal waste, M. van de Voorde ? This is a possible subject for discussion but I first I should like to ask Dr Malášek whether he wishes to add anything regarding fluidity.

E. MALASEK, Czechoslovakia

Generally, we have accepted a slightly different approach to the disposal of bituminous products : below a certain activity, that means to a level in the order of 0.1 Ci/kg,we are using only final burial. We are not intending to temporarily store these materials. For these conditions, we consider only the bitumen itself, without the properties of the containers. That means, for example, in Czechoslovakia we are not even using a metallic drum, but only paper drums which are used in the chemical industry, because this material is satisfactory to provide safe transportation of the bituminous material from the place of origin to the place of burial. The same policy is used in other socialist countries : in the Soviet Union, they are still working on a larger scale ; there were some experiments on the transportation of the bituminous products in liquid form to the waste disposal area, where it is poured into big concreted storage areas or directly into the ground. When the activity is higher, in the order from 1 to 10 Ci/kg of product, the final decision has not yet been taken ; we usually use a metallic container made of mild steel and the products are disposed in concrete trenches. This disposal is also considered as a final burial, but there is a possibility of retrieving the material sometime in the future if it becomes necessary. Our belief at present is that it will not be necessary because the waste disposal areas are selected from the geological point of view with some absorbtion capacity and other requirements ; so it should be safe in the future.

N. FERNANDEZ, France

I should like to comment here. I think that fluidity matters in two specific cases. In interim storage, i.e. where retrieval is necessary, the duration of interim storage should be related to the quality of the container used for the bitumen or vice versa. My second point concerns final storage : there are problems with alpha emitters for which no one here can guarantee that bituminization will be sufficient (in view of the radioactive period of plutonium,which is 24,000 years) and also with the retrieval of this waste, which future generations may have to carry out. Fluidity should therefore be considered in the special context of alpha emitters. I think that it is of little consequence for the other beta-gamma emitters.

H. CHRISTENSEN, Sweden

We have heard very much about bituminization of the different kinds of salt solutions during the Seminar, but we have not heard anything about bituminization of the most common kind of waste which is produced, in BWR or in PWR, that is, different kinds of ion exchangers. Are some of the gentlemen on the Panel ready to say anything about this ?

G. LEFILLATRE, France

I thought that I had briefly mentioned this in my paper on our experiment concerning the embedding and solidifying of organic resins. We have some experience of this problem, which arises in BWRs as well as in PWRs. There does not seem to be any very real problems in solidifying this type of waste using continuous processes, especially under a technique, which I have described, for the direct bituminization of these resins suspended in de-mineralized water in a very dilute (about 10g/l) or very concentrated suspension (200 to 300g/l approximately). I am not referring to discontinuous processes which all have very short contact times whatever the machinery used. Considerable decomposition of organic resins, which are heat-labile products, should be avoided and total dessication should be ensured, i.e. in practice less than 1 % water should be present in bitumen blocks.

This is a technological problem because water is retained to some extent and the same problem is encountered with filters : because of their specific surface they hold water much more strongly than for soluble salts. More or less the same applies to colloidal slurries. At the Barsebäck power station, resin suspensions in active form have been bituminized for the past year. Mr Edwall will be able to reply to your question much better than I.

Y. SOUSSELIER, France

Does anyone wish to say anything about the matter raised by M. van de Voorde on the direct embedding of metals using pure bitumen ?

N. FERNANDEZ, France

Of course, this is something which everyone has wondered or is wondering about. It is very difficult to bituminize metals or solids properly as water may be present and bitumen is cast hot. This poses operational problems : pockets of water may form and bituminization is difficult to control. It would be necessary to dry the material beforehand and to operate at a high temperature, that at which bitumen melts. In any case, the use of concrete here is quite correct. Very good reinforced concrete can be made with steel at relatively low cost and this returns us to the overall economics and optimization of waste management.

G. LEFILLATRE, France

In the context of fluidity, we have difficulties with containers and I should like to mention two aspects of this problem because two separate approaches are involved : the one we and Eurochemic have adopted whereby large interim stocks are set up - in terms of fifty years for Eurochemic and ten years for us - and the point of view held by certain electricity producers such as "Electricité de France", which wish to send their products as fast as possible straight to final storage as soon as they are solidified. At this stage a transport problem arises because some waste can reach 1 Ci/l and shielding then becomes necessary. Shielding involves lead or heavy concrete bunkers and considerable infrastructure and handling equipment is required for transporting the waste to the final storage site. A choice therefore has to be made. This problem arose at Saclay, where it was not wished to keep the waste on the site. It is in fact transported in shielded containers. Retrieval of these shielded containers and permanent rotation was the solution adopted, and the right one as I see it, since one of the chief advantages of bitumen was retained, i.e. volume reduction.

Another possibility is to cast the waste once and for all, at the exit of any bituminization plant, into heavy concrete containers which would offer varying protection in relation to the specific activity of the waste concerned, and then to consider the whole system as waste in its final form. If such a solution is adopted, fluidity will cease to be a problem.

N. FERNANDEZ, France

I am returning once more to the question of final storage, and in particular to the example submitted by the Federal Republic of Germany, i.e. bulk storage in salt mines. Here again the various stages in waste management interact. Although in this specific instance in the Federal Republic of Germany interactions are favourable where bituminization is combined with safe storage in salt mines, this does not always necessarily apply. Bulk storage is not a good example because the third barrier has to be considered if retrieval becomes necessary and I still have alpha emitters in mind.

Y. SOUSSELIER, France

I propose to close the discussion on fluidity at this juncture and to go on to my second question on product quality. The advantage or possibility of separating certain elements in order to improve product quality was mentioned on several occasions. A so-called double bituminization process was also envisaged, as it is possible to apply an outer coating of pure bitumen. This leads us to the wider problem of determining how far research in product quality improvement can be extended.

W. HILD, F.R. of Germany

As far as the characteristics of the products are concerned I think they are, as in the first case which we just discussed, strongly related to the final disposal areas where you will dispose of your material. So any characteristics, resistance to leaching, radiation stability and thermal stability, all these qualities of the product, have to be seen in relation to the final storage area where you are going to dispose of it. I believe that we have sufficient experience to cope even with material which from the first point of view does not completely satisfy our postulations derived from our final storage site. We can act as already mentioned on chemical compositions, as for instance in the case of Eurochemic a chemical treatment procedure can be provided to come to a product which at the same time is more resistant to corrosion and more resistant to leaching. You have the problem of applying a cover, in one case it is the containment, in the other case it is an additional layer of bitumen. We have sufficient means in my opinion - and I throw this open to the floor - to act and to play on the quality as it is postulated. I am stressing again that our postulations for the product characteristics are mainly determined from the conditions in our disposal area. In this respect, we have also to see the other characteristics ; the thermal stability of products incorporated in the bitumen matrix and the amount of radioactive nuclides, alpha or beta-gamma emitters, as far as this is in connection with the radiation stability.

N. VAN DE VOORDE, Belgium

When for instance, M. Lefillatre was saying that ion exchangers should not be decomposed but preserved as such in bitumen, as also sodium nitrate, borates, etc., which are not radioactive, I wondered whether we should examine the problem of medium level waste, which we have all helped to create. Could we perhaps tackle the problem of producing this type of waste ? Is it completely impossible to avoid bituminizing ion exchangers ? Is the bituminization of soluble salts unavoidable ? There are other ways of dealing with this problem. Cannot ion exchangers be regenerated and waste precipitated from the regeneration solutions and only high level waste directly vitrified, and what about low level waste which every one agrees should be bituminized ? Fluidity would no longer matter. Bituminization would be valid simply for low activity waste. Vitrification would be required for high level waste. As for the treatment of medium levels, we shall have to study this in future.

E. ESCHRICH, Eurochemic

We have already joined the army to fight in this problem area. We can distinguish between low level waste, medium level waste, between short lived waste and long lived waste, and we can imagine combinations of all of them. Whatever we do, the waste management problem is finally a storage or a disposal problem. We condition the wastes to ensure safer isolation at the storage place, so that we really achieve the ultimate aim of all waste management activities which is to avoid hazardous doses of radiation reaching man or other living beings from external or internal radioactive sources. To stress a point made by Dr. Hild, finally we always have to look at the interrelations of product qualities and storage conditions.

If there existed a guaranteed eternally-safe disposal place one would not need any elaborate conditioning at all. If we doubt the safe isolation, adequate measures have to be taken to outweigh any insecurity factor, and if we do not know at all where and when the waste will be finally stored or disposed of, we have to apply the best waste conditioning techniques presently available.

What is normally doubted in long-term security considerations is man's ability to predict situations on earth over time periods of some hundreds of thousands of years and longer. Unpredictable changes of climate, societies, geology and so forth may endanger all that we have done with the intention of safely isolating the wastes over the periods required.

Looking at the long-term risk leads unavoidably to the problem of alpha-emitters in the various wastes. Some types of reactor wastes are free from actinides and do not present a hazard after a relatively short storage time, whilst the wastes deriving from reprocessing operations are always suspect to contain some alpha-emitters.

Normal medium-level waste from reprocessing has not the same composition, as far as alpha-emitters are concerned, as the fission product high-level waste, simply because it originates from other stages of the process. There, we have mainly to deal with plutonium, uranium and some americium from the plutonium 241 decay. It is certainly possible, and we have proven it by laboratory experiments but not yet on an industrial scale, to remove almost quantitatively all the actinides from medium level waste solutions using chromatographic methods. Our tests have furthermore indicated that the removal of fission products from the large excess of inert salts, usually sodium nitrate, is achievable by relatively simple means. However, the removal of alpha emitters to the required degree would already result in an essentially improved bitumen-waste product necessitating only a relatively short storage period - at least one that can be overlooked with certainty - until the contained fission products have decayed to innocuous levels.

W. HILD, F.R. of Germany

I just wanted to add to the remarks of M. van de Voorde and Mr. Eschrich that I am very much in favour of this cutting down of our waste problems into two waste problems : high level waste and low level waste problems. As many of my former Eurochemic friends are here, they all know we have been discussing this for a long time. As a matter of fact, we have at Karlsruhe also finally started a programme that aims at separation in these two directions, high level waste and low level waste. Splitting the medium level waste into these two categories would be very interesting and very wise to do in the future. However, we should always stress the point that waste management should start at the origin where wastes are produced. In all nuclear installations, both designers and operators should be aware that they are producing waste, and they should do the utmost possible to design processes in such a way that the waste production is kept to as low a volume as possible, and to operate them in such a way that as much as possible of the material can be recycled. If we have all these combined efforts coming from the part of waste managers and all those who are running and operating power reactors and reprocessing stations, I think we have in the future a good probability to find new and even more effective processes for waste treatment and disposal.

E. MALASEK, Czechoslovakia

Much has already been said about the influence of disposal conditions on the production of bituminous products. However, there is one more factor which has not yet been discussed, that is our tendency to incorporate as much as possible of radioactive concentrates into the minimum amount of the final product. This again may influence the conditions of disposal. In some cases, there is another additional factor, that is the requirements of the process itself : in some combinations of salts or sludges, it is possible to have much higher concentrations from the storage point of view, but the temperature or the process which is selected for the treatment of the material, or even the bitumen itself are not suitable to incorporate higher amounts of this material. Another way to improve product properties is to minimise the amount of soluble material which is included in the bituminized products. It would agree with M. van de Voorde as to his question concerning the bituminization of ion exchange material. We have done a great deal of work in this field and I would agree that it is possible to bituminize ion exchange materials, but we are not yet satisfied with the properties of such products, specially for long term storage. At present we are not really able to give any definite answer and our tendency is to minimise the volume of ion exchange materials which is necessary to dispose of. I would personally prefer the method which was published in Vienna by M. van de Voorde concerning the incineration of this material before bituminization.

G. LEFILLATRE, France

I would like to make three comments. The first is on the quality of the finished product in relation to storage. Obviously, before a solidifying process is defined, the nature of the final storage to be used for this type of waste should be known. Nevertheless, irrespective of the type and location of final storage, transport will always be a problem. It seems to me out of the question not to envisage waste solidification if only to ensure transport safety between the production site and final storage site. Consequently, if bituminization is a viable process it may fulfil these conditions as do other processes, since here again it has the advantage of reducing the volume of the waste. Along the same lines, in the case of a very safe form of storage such as salt mines, there is no point in studying ways of improving this waste as a high safety level has already been reached at final storage. Transport safety is the only remaining problem.

My second remark concerns resins, and I entirely agree with M. van de Voorde's comments. Unfortunately I note that the problem of spent resins is quite typical of light water reactors. In reprocessing plants, we use ion exchanger resins to treat much higher level waste than in reactors, or at least of the same order of magnitude in certain fields, and so far we have never had any problems with spent resins because they are systematically regenerated (this applies to France, both at Marcoule and at La Hague). This matter should be discussed with electricity producers as it is perhaps also an economic problem, since the cost of ion exchanger resins is constantly increasing and the volume of resin losses is very high.

My third comment, still on M. van de Voorde's remarks, concerns the bituminization of salts which are in no way radioactive. I should like to point out that his remarks do not follow the present trend of "zero discharge", because in that case highly complex chemical processes will be required and furthermore, there will never be any certainty of obtaining totally inactive salts. The problem of the final embedding of these salts will therefore always remain. Naturally, separation would be desirable for alpha emitters, in order to bituminize maximum level waste in a reduced volume, and such a measure would be positive. Finally, even on the assumption that very high decontamination factors are present, there will always be low or very low level waste which cannot be discharged into the environment without prior embedding.

N. FERNANDEZ, France

I think that the standard of the embedding process matters because it enables certain complex problems such as the containment of ion exchanger resins, heavy solvents like TLA, etc. to be resolved. If the study of the quality of these bitumen blocks is to be pursued, it should be with the purpose of resolving the problems which may arise and which should always be taken into account, because there will always be operational abnormalities. I know that M. van de Voorde, Mr. Hild and Mr. Malášek are basically right, but there will always be operational abnormalities, medium level waste problems, since very often no overall treatment system is available under which it is possible to move from one category of level to the next. Furthermore, perhaps one aspect of quality has not yet been raised, that of monitoring the quality of the product emerging from the plant. I do not think that such monitoring should be carried out in relation to the optimum quality attainable under this process, but in relation to the average quality required to ensure future storage safety. The whole procedure must lead to relatively cheap monitoring. If control plants as sophisticated and advanced as those of Eurochemic had to be set up everywhere, it would be frightfully expensive. We have come up against the economic aspect again, and a relatively cheap monitoring system would have to be adopted in order to obtain an adequate average quality.

J. ORTEGA ABELLAN, Spain

I believe that the conversion of nitrates to nitrites is a problem. Nitrates may react with the hydrogen released in radiolysis to produce nitrites, which constitute a safety risk. I think that in time, large quantities of nitrites might be formed.

G. LEFILLATRE, France

A few years ago, around 1965-1966, the Americans already examined the problem of nitrite formation in solutions of fission products, following the development of reprocessing centres and the increase in reactor power.

According to research done at Oak Ridge, very serious problems were encountered with mixtures of bitumen and sodium nitrite in certain proportions. We repeated systematic tests with widely differing mixtures, also including very low amounts of bitumen in order to take into account problems of scale formation in equipment. For example, at 90 % salt, a rather utopic level, and 10 % bitumen, no special exothermic problems were ever observed. The troubles I referred to in my paper were mostly due to ammonium nitrate and personally, I do not think that sodium nitrite is a cause for concern.

W. HILD, F.R. of Germany

I should like to stress the point which M. Lefillatre has just made. We have more or less the same experience as he has and I believe the nitrite problem is only to be considered as serious if you look into really high level waste, so that you have high doses of radiation causing this radiolytical destruction to nitrites. What is dangerous, and this is well known by all of us, is if we have a system of ammonium nitrite, because this is really easily decomposed and can even act as an explosive ; this therefore has to be excluded, but there are sufficient means to exclude ammonia. Apart from that, as far as medium level waste and low level wastes are concerned, I do not consider it as a great problem.

H. ESCHRICH, Eurochemic

I just want to add that the amount or the concentration of nitrite you might fear to have in the solution which would finally end up in the bitumen product, very much depends on how you store your liquid waste. If it is stored in an acidic medium, then the nitrous acid has a low stability and solubility ; the equilibrium concentrations I do not know exactly by heart. They depend naturally on the total free acidity and the temperature, but they certainly do not exceed some hundredth of a molar in aqueous solutions. If the solutions are stored in an alkaline medium one would risk the build-up of nitrite. On the other hand, if you fear that there might be problems connected with the presence of nitrite which we have just heard are very rare, it is very simple to destroy it, either just by simply sparging acidic or acidified solutions or by adding reagents which react quantitatively and rapidly with nitrite and lead to products which are completely innocuous, like nitrogen and water for example. I do not think therefore that nitrites constitute a problem which could cause major difficulties. Those can easily be overcome if one fears that certain nitrite containing solutions may create a hazard when subjected to bituminization.

R. SIMON, C.E.C.

Could M. Lefillatre supply any figures on the costs of bituminizing the residues of boiling water reactors, such as amberlite or microsphere resins ?

G. LEFILLATRE, France

It is difficult to give a price off the cuff as many factors are involved. At CEA, prices are of the order of FF 900/m^3, including transport, the collection of effluents at the Centre, evaporator concentration, bituminization and subsequent transport to the storage site. To give you some idea of the order of magnitude, the cost for Marcoule is about FF 10,000/m^3 of defined waste. This compares well with the figures given by Eurochemic yesterday, which were approximately FF 12,000/m^3.

Y. SOUSSELIER, France

If there are no further questions on product quality, can we go on to the third topic I proposed to you, that of maximum risk, or in other words the fire risk. This morning we heard that there was no need to adopt an alarmist viewpoint here. Nevertheless bitumen can burn and this may occur during manufacturing operations, when it would appear easy to control, during transport, when once again it might be fairly easy to control, and above all during interim storage and this is what I should like us to discuss.

In the case of interim storage for several decades, as at Eurochemic, should this risk be taken into account, and can product quality be improved even if it means additional costs ? It is possible to avoid taking steps such as the use of extremely elaborate storage bunkers for protection against air crasches ?

H. ESCHRICH, Eurochemic

The possibility that the active bitumen-waste product may burn has always been considered as one of the main disadvantages of bituminization and it is the main reason that some people prefer other waste treatment techniques. The very few incidents involving burning of bitumen products have up to now always happened during or directly after the incorporation process itself. We now know that we have enough remedies to eliminate the risk of having a fire at the critical spots in the installations. But the risk of having a fire during interim storage also exists and various measures can be taken to reduce this risk to an acceptable minimum. One of them could consist of artificial burying within an engineered storage containment, for example by filling the voids between the drums or other packages with sand or another non-burnable material, which, when it is necessary to retrieve the waste, could be easily sucked out, especially if one is sure that there are no transferable radionuclides or loose matters in the fire-protecting material. This measure is possibly a littre exaggerated. Our view at Eurochemic is the following : in a fire there is first something which can burn and secondly another which makes it burn. The thing which can burn is the waste ; we aim at eliminating what gets it burning. This is our approach to interim storage in our bunkers : there will remain no installations or equipment in the bunker after we have filled it which could give rise to sparks or ignition. We have no lighting, no electrical installation or apparatus, nothing which could give sparks, nothing which moves. Static electricity is normally absent in an atmosphere that is completely ionised which is certainly the case in the radiation field existing in the storage bunker. Forced ventilation prevents any accumulation of explosive gas mixtures. We do not believe therefore that we have the risk of a fire as far as the entire storage system within the containment is concerned. Precautions have been taken to enable the cooling of the bunker surfaces by water from the outside and to fight a fire approaching from outside the storage rooms. If bituminized waste burns there is normally a gas phase and an ash which is left. The burning of bitumen products has been investigated and some of the experiments have been done at GfK in Karlsruhe : in the order of 20 to 30 % of the incorporated mixtures of fission products were "volatilized", while the rest of the material was found in the residual ash. Therefore, even a fire, especially in a containment which is dimensioned and built with such a risk in mind, will not necessarily result in the spread of radioactivity to the environment, but I repeat that for our interim storage we have primarily taken measures to eliminate the causes of ignition of the products stores rather than to fight the fire.

W. HILD, F.R. of Germany

I would like to state once more that we all are aware that bitumen and bitumen products are combustible materials ; but we are also aware that the material is not easily flamable and there are sufficient means of protecting the material, even during intermediate storage to avoid the material from catching fire. As I showed you in my presentation, an extremely hot source is needed to get a block of bitumen burning. This is why we are rather confident, particularly if you look at the possibilities existing to counteract a fire, and to control storage areas.

Y. SOUSSELIER, France

I would like to ask M. Fernandez's opinion on this point as he is directly involved in these storage problems.

N. FERNANDEZ, France

I agree with Mr. Hild that it is very difficult to make bitumen burn. It required a very high energy source. In fact a very large dimension accident is involved, i.e. the impact of an aircraft filled with kerosene which would ignite the stocks by providing a large heat source. This is probably extremely unlikely to happen, but it still has to be considered.

Y. SOUSSELIER, France

In other words, do you think that it ought to be taken into account ?

N. FERNANDEZ, France

If the local safety authorities in a country say that such accidents must be taken into account, then they must be taken into account.

R. SIMON, C.E.C.

Mr. Eschrich has mentioned that a certain fraction of fission products will escape as volatile material in case of a fire. What will happen to this volatile fraction ?

W. HILD, F.R. of Germany

Mr. Eschrich has mentioned experiments which we performed together with the Institute of the chemistry of propellants and explosives. Part of the slides I showed during my presentation were taken during these investigations. They showed that we carried out tests for various firefighting measures. We also wanted to get an idea of the spread out of radioactivity if you have a product without any action on the fire. Without firefighting, we kept a product with a high nitrate content burning until the end (a 175 litre drum filled with approximately 50 weight % bitumen sodium nitrate mixture). To get an idea about the spread out in the surroundings, we collected the material which came down with the fumes, we measured the wind and the values Mr. Eschrich cited are in the right order of magnitude. We analysed the material which was contained in the ashes of the drum and the rest of the drum, counted it and made a balance for the material which was found in the various dishes standing aroung and collected. Due to the fact that we had quite high sodium nitrate content, we believe (and this is a result of the analysis done) that this volatilisation is not a volatilisation in the proper sense : it is more or less a carryover of sodium hydroxide dusts and aerosols. In fact it is in the order of 20 %, but if you are interested we can give you the exact figures.

K.A. TALLBERG, Norway

A question on this fire experiment : Did you continue to add gasoline to make it continue to burn or was it self-burning all the time ?

W. HILD, F.R. of Germany

In this particular experiment, we utilised our experience from the first combustion experiment. We added a sufficient amount of oil to start the fire and to get it to a point where it was self-sustaining.

Y. SOUSSELIER, France

Perhaps we could follow on to the next point, which we have already briefly touched upon : overall optimization of the process and whether we have now reached a stage where the overall management and procedure of bituminization are optimized, assuming that this management can be regarded as extending to the final storage stage. If final storage cannot be taken into account, interim storage can be regarded as included.

My question covers in particular the problems relating to the separation of certain products, which would, for instance, entail additional costs but would perhaps subsequently lead to lower storage costs or lower safety costs under less stringent safety authority requirements. My question is difficult as it leads us to the topic of costs which are never very easy to estimate, but I nevertheless do think that we should broach this.

G. LEFILLATRE, France

To my mind, the chief problem in future studies on the bituminization of radioactive waste lies in the following : what is a reasonable limit for the bituminization of radioactive waste ? Today, speakers referred to 1 Ci/l as a current reasonable limit. A few years ago, a figure of 10 Ci/l was quoted. In my opinion, no one has the faintest idea. This is therefore the main problem outstanding.

The second is that of alpha emitters : what limit could also be set for the bituminization of this type of waste so as to obtain good safeguards in long term storage ?

These are the two major types of study to be undertaken in the near future in the bituminization of radioactive waste.

Y. SOUSSELIER, France

I shall still ask you another question : how do you think this research should be guided as these are very general research topics ? For instance, what programme could be conducted in order to determine the limits, up to 1 Ci/l or 10 Ci/l, provided that the studies are still carried out fairly rapidly in view of the development of nuclear energy and bearing in mind that in certain instances, long term behaviour is the problem ?

G. LEFILLATRE, France

Owing to delays and costs, the French programme has slowed down at the moment but next year we hope to be able to resume an experiment on the bituminization of fission products at 100 Ci/l, which was stopped in 1971. Naturally we do not envisage bituminizing products at 100 Ci/l in the future but we wish to obtain an increase factor of 10 at least in order to accelerate the self-irradiation process and to be able to calculate more accurately than at present, in my opinion, the release of radiolysis gases and heating problems, especially with gamma emitters. In the alpha sector, we also have an actinide separation programme, either by means of ion exchangers or by chemical methods. On the basis of this separation, we subsequently intend to bituminize alpha emitters with maximum specific activity (plutonium, curium, americium), in much the same way as our colleagues have approached vitrification, provided that we do not encounter criticality problems.

W. HILD, F.R. of Germany

I think we should continue with our investigations on radiation stability. M. Lefillatre noted that it would be interesting to continue with these experiments and our colleagues from the Eastern countries have started some investigations in that field. We need still some further clarifying points as far as the comparison of external irradiation tests and internal irradiation tests is concerned. In this connection what is very interesting - and I think a very good contribution to this problem is the proposed foreseen experiments at Eurochemic together with our collaborators from Karlsruhe - is to try to look into the stability of bitumen products in relation to higher concentrations of typical alpha emitters. We could perhaps foresee a similar experiment as is actually being performed on vitrified products : a kind of a time lapse experiment where you incorporate curium 242 in the bitumen product, to get in a rather short time the possible radiation damage a bitumen product would receive if stored for a time period corresponding to the complete decay of the actinides present. These are worthwhile questions to look at.

H. ESCHRICH, Eurochemic

My contribution will be very theoretical because we do not exactly know what the operation costs will be in our installations as we have not had active operation experience up till now. However, an installation similar to the Eurochemic one is not always necessary to treat nuclear wastes. For example, I can imagine that a bituminization plant at a reactor site can be much simpler in its set-up and also most probably cheaper in investment costs. Besides this, the manpower required to get the installation properly running depends on how many controls, how much automatism one wants to add or not. Furthermore the product quality can also be influenced by available financial means : for example, we have already mentioned that one can add very costly reagents ; but one can also use cheaper ones which may diminish slightly the product quality, but to an acceptable degree. One must not necessarily go to the extreme in achieving the best possible product. If one wants to do this, one could do something, for instance, against the combustibility as was tried some years ago in the United States ; there exist additives to reduce inflammability and to retard the burning and the burning rate. As far as the maximum allowable specific activity is concerned, there are still discussions going on and we will produce and investigate some industrial size products of high specific activities so that we can contribute, by results obtained, on a real industrial scale to clarifying this problem.

From the theoretical point of view, if a bitumen type is assumed to be radiation stable to say 10^9 rad, it is possible to calculate the integrated dose a bitumen matrix, having incorporated 1 Ci/l of mixed fission products, will receive, and to find out that 10^9 rad are not accumulated until complete decay. On theoretical grounds, this would be equivalent to about 100 Curies per litre, or somewhat less. However, this is theory and we have to confirm it by experiments.

E. MALASEK, Czechoslovakia

I am not able to give any absolute figures about the economy of bituminization processes. We have done such work in evaluating the economy of the complete waste management system, when we changed the previous cementation process and disposal area to a bituminization process and a new disposal area. The result was the decrease of the total cost to 60-80 % in comparison with cementation, but the main factors were in this case the cost of one cubic metre of storage area, and then the question of containers for bitumen in comparison with cement. Generally, it has been proved that in comparison with cementation the bituminization process is more economic. As fas as the maximum activity which can be incorporated into bitumen is concerned, at present our problem is that up to the level of one Curie per litre any bitumen will be suitable with no special requirements for removal of gaseous products. For higher activities, we are storing materials with specific activities in the order of 10 to 50 Curies per kilogramme of product. For these cases hard type of bitumen is used ; the tendency is to get a porous material without swelling with the possibility to free gaseous products which are removed from the storage area during an initial period of time. It is considered that after some time, depending on the activity, it would be possible to stop the ventilation system and to store without ventilation.

N. FERNANDEZ, France

As an operator, I feel I have a say in the matter. Optimization in waste management is always limited to the same problem, that of the risk presented by the waste in relation to the process used. As far as we are concerned, it is bitumen, with its special drawbacks. However this general problem can be split up into several sub-problems, including the chemical engineering of the plants, especially interventions and maintenance. Plants should be designed so that maintenance is easy and the minimum of secondary waste is produced. Monitoring of product quality is also involved and I think that the minimum monitoring required for guaranteeing average quality of the product obtained should be carried out. Potential risk is also a problem, especially with alpha emitters. I feel producers should bear the additional cost linked to the type of final storage required. This might encourage producers not to produce this type of waste. Then we also have the problem of general restraint, not only on the part of producers but also on that of high level officials who must impose this restraint and on that of others who have to comply with it. It is a matter of responsibility. If these objectives could be met the economics of the process could obviously be optimized. In any case waste management will always be relatively expensive.

Y. SOUSSELIER, France

Everyone will heartily agree with what you have just said. More generally, it seems to me that in recent years we have discovered that waste management is more expensive than we had originally anticipated. Although increases in this sector may have been lower than in other nuclear energy sectors, they have been by no means negligible

N. FERNANDEZ, France

I am a little disturbed because the more we discuss processes the more we look for very complex and over-elaborate processes which be increasingly expensive.

P.W. KNUTTI, Switzerland

Mr. Eschrich mentioned that some additives exist to reduce the combustibility of bitumen. I am wondering if this is a technique which could be applied on a full scale scheme and what additives they are. Could Mr. Eschrich elaborate on this possibility, because if this exists, we have found a beautiful product to treat our wastes ?

H. ESCHRICH, Eurochemic

These additives are sometimes called "fillers". The incorporation of these additives does not mean that bitumen becomes unburnable ; il still remains a burnable product but what one achieves is fire retardation. What I am thinking of are fire inhibitors like aluminium hydroxide or phosphates which we have now used in one of our pretreatment flowsheets. The more one adds of these additives the less bitumen becomes inflammable. You then need quite a long sustaining flame to get it to start burning. Naturally if one can reduce the oxygen containing compounds, then products can be obtained of a better quality than we normally get from the medium level waste from reprocessing plants, for example. We have to make a careful cost/benefit analysis to decide whether substances which delay fire should be incorporated in the bitumen product. One should be aware that conditioning of waste by even an elaborate bituminization technique including on-site handling and interim storage requires only about some tenths of the total waste management costs, most of the costs being due normally to transport and final storage. This is an estimate as nobody can accurately specify these costs, but it is approximately the distribution one can foresee. You can therefore use suitable additives to improve your product if you do not trust the safety features of the interim or final storage places too much.

P.W. KNUTTI, Suisse

How does this affect the other material properties such as leachability ? Has it a negative effect ?

H. ESCHRICH, Eurochemic

If solid fire inhibitors are added, either the percentage contents of the solids in the final product or the volume of the final product will be increased. If a high percentage has to be added to achieve the desired effect one may expect that the leach rate of certain nuclides can negatively be influenced as the available bitumen coating layer of the solids becomes smaller.

However, one can also add materials which lower combustibility as well as the leachability of the products by acting in addition as scavengers or sorbents. Whether the additives in question will affect negatively or positively the final product properties depends on their nature, the amount added and their distribution within the bitumen matrix. Properly chosen they will not weaken those properties of the products essential for safe long-term storage.

Finally we always have to look at the combination of the quality of the bitumen-waste product and the quality of the storage both together, to obtain the required degree of safety.

Y. SOUSSELIER, France

It is always a matter of optimization ! Any other questions ?

E. BREGULLA, F.R. of Germany

To the question of economy, I would like to make two comments. First, looking around in different laboratories and other institutes, use of large quantities of water to reduce the activity looks quite similar to the use of salts and other detergents. Second, we have to give more attention to the design and construction of equipment for maintenance and replacement purposes, because the money and time we save in this phase are spent later twice and even more. It affects shut-down times, and produces man rem too.

Y. SOUSSELIER, France

We still have to discuss the choice of bitumen and whether any research is required on the various grades of bitumen, particularly in relation to the chemical nature of the petroleum raw material. Is there any point in undertaking this research, possibly under international co-operation since each of our countries only has a limited number of crude petroleum sources and we know that certain crudes have very different characteristics ?

G. LEFILLATRE, France

The main difficulty in selecting bitumen is precisely the great variety of crude petroleum sources. Within the present context of the energy crisis, petroleum companies are no longer able to guarantee a specific source of petroleum as they could ten years ago. Very wide variations in the physico-chemical properties of the bitumens supplied may therefore occur. This is troublesome for the highest activity waste especially, i.e. the category above 1 Ci/l mentioned by Mr. Malášek. I think that this category should be studied. Road surfacing is carried out with materials which no longer bear much relation to bitumen but are sulphur derivatives with better ageing properties than the traditionally used bitumen. These materials could be investigated with a view to manufacturing a special type of bitumen, possibly a "vulcanized" bitumen, by means of organic chemistry studies with the collaboration of oil companies, in order to check whether this type of bitumen would be appreciably more resistant to radiation than those used at present. If such was the case, manufacturing consistency would be feasible as a special bitumen for atomic use would be produced.

I have always been puzzled by the fact that petroleum companies do not know a great deal about bitumen, except that it is a by-product of the petroleum industry, and that they have never carried out any in-depth theoretical studies on bitumens as they have on polyethylenes and chemically well-defined products for instance. This would therefore require substantial investments, as theoretical studies would be needed. So far, to my knowledge little had been done in this field anywhere in the world except, perhaps in the USSR, and there is scope for study.

E. MALASEK, Czechoslovakia

At present we are considering in the first place the economic question when using bitumen. We are trying to use the bitumen which is normally available in the country and what we have done is to verify the properties of these materials. The comparison has shown that what is more decisive is not the source of bitumen itself, that means the source of petroleum, but the method of treatment of petroleum. So in our conditions, when the source has been changed the change of properties of bitumen was relatively small and it was possible to use the product of the same plant for bituminization of wastes without any special difficulties. Of course, it is again necessary to divide the low activity and the higher activity waste products. What is quite appropriate for activities at the level of 1/10 Ci/kg may be unsuitable for higher activities. From this point of view we consider it necessary to verify the properties of any new type of bitumen which appears in the future and to put it in some category comparable with the material we have used already. I do not think that at present, in our country, it will be reasonable to produce some bituminous products only for insolubilisation of wastes, because here again it is a question of economics. We can get our bitumen partially as a waste product and at a relatively low price. Asking our industry to develop a special material for incorporation of wastes would increase the price 100 times or more. It is better to use the material which we have and not to ask such a question to the industry.

W. HILD, F.R. of Germany

As far as the bitumen grades selected for particular incorporation operations are concerned, we have all developed processes for incorporation, starting with investigations to select the proper bitumen that is suitable for our particular case. Here again, there is an interaction between product characteristics and final storage for instance. All of us have performed experiments in this respect and have gained quite good experience as far as these interactions are concerned. Starting from there, we now have in various countries a number of technical installations running for a certain time with different grades of bitumen. There is sufficient experience available, and my personal opinion is for the time being that we are in a rather good situation ; we know a great deal about the bitumen grades that we have to select and I do not see an immediate need for actions to find better bitumen material. I think the data we have available are quite sufficient for the time being.

H. ESCHRICH, Eurochemic

The fact that the petrol firms do not make a special effort concerning our problem shows that we are not a big consumer !

Y. SOUSSELIER, France

I think that you have made a very pertinent remark ! Before closing our discussion, need I add further conclusions than those which I proposed when opening the session ? I believe that this Seminar has really enabled us to show that bituminization is a process which works and which produces good results. Naturally we have seen that a number of uncertainties subsist although none is likely to call the process into question, and I think that from this angle, we may continue our endeavours without hesitation.

Ladies and Gentlemen, I hereby declare the Seminar on bituminization closed.

Y. SOUSSELIER, France

On pouvait se demander, avant l'ouverture de ce Séminaire sur le conditionnement dans le bitume des déchets radioactifs de faible et de moyenne activité, si une telle réunion serait vraiment très intéressante, puisqu'elle venait moins d'un mois après le très important Symposium organisé conjointement à Vienne par l'AIEA et l'AEN sur l'ensemble du problème des déchets et où, bien entendu, le problème du bitume était déjà traité. On pouvait aussi se demander à priori si un tel Séminaire, ne regroupant que des spécialistes, ne se traduirait pas par une sorte de congratulation mutuelle ou de satisfaction exprimée de part et d'autre. Il est évident que rien de tout cela n'a eu lieu, les discussions que nous avons eues sur les très intéressantes communications qui nous ont été présentées et la visite très complète des installations d'Eurochemic nous ayant appris et apporté beaucoup aux uns et aux autres. Le premier enseignement à en tirer, et que certains avaient déjà tiré à la suite du Symposium de Vienne, est qu'une réunion est plus fructueuse si elle est limitée à un nombre restreint de spécialistes qui parlent de sujets qu'ils connaissent bien et qui peuvent, par conséquent, en discuter de manière approfondie.

Il est possible, d'ores et déjà, de tirer un certain nombre de conclusions ou plutôt un certain nombre de confirmations. L'enrobage par le bitume des déchets de faible et de moyenne activité est un procédé qui est au point. Il y a un certain nombre d'installations en fonctionnement dans le monde ; les résultats obtenus jusqu'à ce jour sont extrêmement satisfaisants ; les incidents ont somme toute été très peu nombreux et pour les plus importants d'entre eux (qui n'ont pas eu de conséquences importantes) on sait très bien quelle en a été la cause et comment les éviter à l'avenir. L'expérience dont on dispose en matière de stockage des produits obtenus est également satisfaisante.

Cela veut-il dire que le travail est terminé, qu'il n'y a plus simplement qu'à construire d'autres installations et les faire fonctionner pour le plus grand bien de l'énergie nucléaire ? Certes, cela est vrai, mais nos discussions ont quand même bien montré qu'il y avait encore un certain nombre de problèmes qui se posaient. J'en retiens certains, notamment sur le fluage du bitume et les problèmes que cela pourrait par exemple entraîner dans un stockage comme celui réalisé actuellement par Eurochemic, s'il faut au bout de plusieurs décennies reprendre ces conteneurs alors que les fûts auront été plus ou moins attaqués, ou auront plus ou moins disparu à cause de la corrosion. Une autre question avait trait aux risques d'incendie. Nous savons que ce problème a été longuement débattu dans certains pays et il est probable même que si certaines nations, comme les Etats-Unis d'Amérique, ne se sont pas lancées plus rapidement dans des études sur le bitumage, c'est parce qu'elles ont attaché à ce risque une importance assez considérable. Au cours de la visite que nous avons faite à Eurochemic, j'ai entendu poser des questions sur le risque maximum pour le stockage, lié par exemple au problème de la chute d'avion sur le stockage. Une autre question était de savoir s'il était intéressant d'avoir un produit aussi élaboré que le bitume quand on disposait de possibilités de stockage définitif offrant à elles seules une garantie très élaborée. D'autres questions avaient trait au choix de la qualité du bitume, et l'on n'a peut-être pas assez insisté au cours de ce Séminaire, probablement parce que tout le monde ici en est conscient, sur le fait que le mot "bitume" recouvre une catégorie très large de produits qui diffèrent notablement par leurs caractéristiques physiques, chimiques ou physico-chimiques.

Bien entendu, toutes ces questions sont liées, ainsi que l'ont montré les discussions. D'une façon générale, pour la gestion des déchets, on doit considérer que depuis la production jusqu'au stockage final, il y a interaction entre les différents stades. Si l'on veut optimiser un procédé, voire même choisir un procédé, il ne faut pas se contenter d'examiner ce qui se passe à un point bien défini et faire de l'économie dans un domaine limité ; il convient au contraire, d'une façon beaucoup plus générale, de regarder quel est l'ensemble du bilan économique, l'ensemble de l'analyse coût/bénéfice pour la gestion complète des déchets depuis la production jusqu'au stockage final. C'est bien sûr une tâche difficile, souvent même une tâche impossible actuellement, parce que dans beaucoup de nos pays le mode de stockage définitif n'est pas encore absolument défini. A dire vrai, seuls nos collègues allemands ont, me semble-t-il, défini d'une façon bien précise quel sera leur mode de stockage définitif. Dans la plupart des autres pays, il y a encore un certain nombre d'options ouvertes, un certain nombre d'alternatives possibles.

Néanmoins, et sans vouloir discuter dans leur ensemble de tous ces problèmes d'interactions et de stratégies de gestion, il me semble que nous pourrions discuter d'un certain nombre des points dont nous venons de parler. Je vous proposerais que nous les classions (c'est toujours un peu arbitraire) en cinq groupes de questions :

- La première pourrait porter par exemple sur les problèmes liés au fluage du bitume. Ces problèmes doivent-ils être considérés comme importants ? Si oui, peut-on envisager des méthodes plus ou moins simples ou élaborées qui palieraient cette difficulté ? Est-ce que ces problèmes de fluage doivent entraîner sur le plan du stockage des décisions comme, par exemple, l'envoi le plus rapidement possible des déchets à un stockage définitif ? Ou bien, faut-il avoir des conteneurs plus élaborés que ne le sont les conteneurs actuels ?

- La deuxième question, qui est aussi une question très générale, a trait au problème de la qualité du produit. On nous a beaucoup parlé de taux de lixiviation ; nous avons vu que ces taux sont très bons et qu'ils sont finalement souvent supérieurs à ceux que l'on avait envisagés au départ. Mais la question peut toujours se poser de savoir s'il faut essayer d'améliorer cette qualité ou au contraire si, à partir du moment où l'on atteint une certaine valeur et compte tenu des radionucléides présents dans les déchets, il n'est pas nécessaire de poursuivre les efforts visant à l'amélioration du taux de lixiviation.

- Le troisième groupe de questions porte sur le problème du risque maximum lié à un incendie au cours d'un stockage ou à un incendie lors d'un transport. Doit-on prendre ce risque en considération et si oui, comment peut-on le palier, que peut-on proposer pour le diminuer, voire le supprimer ?

- Le quatrième groupe de questions a trait au problème difficile de l'économie globale du procédé et de son optimisation.

- Enfin, dernière question relative au choix du bitume et à l'intérêt éventuel d'essayer de déterminer, parmi tous les types de bitumes de toutes origines et de toutes compositions, si le choix a ou n'a pas une grande importance sur le procédé.

Je propose de demander d'abord aux membres de cette Table ronde ce qu'ils pensent de ces questions. Puis ensuite, je demanderai aux membres de l'assemblée s'ils ont des questions à poser, des remarques à formuler. Je souhaiterais qu'au cours de cette Table ronde, on évite les questions trop spécifiques et que l'on donne aux discussions un caractère relativement général, compte tenu des conséquences éventuelles que ces discussions peuvent avoir sur nos programmes de recherche et de développement.

Nous allons donc aborder en premier lieu le problème du fluage. Ce problème est-il important ? Peut-on l'éviter ? Doit-on en tenir compte dans le stockage ? Puisque c'est M. Lefilattre qui a posé le premier la question à propos du stockage à Eurochemic, je voudrais lui demander ce qu'il en pense.

G. LEFILLATRE, France

Si j'ai posé la question, hier, en ce qui concerne le stockage temporaire d'Eurochemic, c'est qu'initialement, ce stockage avait été prévu pour une durée de dix ans et que dans ces conditions le problème de la corrosion des fûts ne se posait pas. Mais ce temps de stockage a été augmenté ensuite pour aller éventuellement jusqu'à cinquante ans. J'ai donc posé cette question en pensant qu'il se pourrait que des fûts en acier doux ordinaire soient corrodés au bout d'une telle période. Il m'a été répondu qu'on avait toute garantie sur ce point à la suite du choix plus sophistiqué d'un acier chromé. Je pense personnellement que ce problème existe au niveau du stockage temporaire, mais qu'il n'existe plus au niveau d'un stockage définitif, ce dont on pourra discuter plus en détail par la suite. Je ne vois pas cependant comment on pourrait empêcher le fluage du bitume, même en ajoutant des charges inertes qui poseraient des problèmes technologiques. Dès l'instant où le bitume est une matière plastique, il fluera inévitablement dans le temps, et il faut étudier ce phénomène du point de vue risque, que ce soit un stockage définitif sur le sol, ou en profondeur, ou au fond de la mer. Enfin, compte tenu des caractéristiques du produit du point de vue de la lixiviation, s'il n'y a pas de problème de température et d'auto-échauffement (c'est un cas qui est à prendre en compte pour des stockages définitifs de déchets qui dépasseraient 1 à 10 Ci/l, surtout 10 Ci/l), je pense que tous mes collègues et moi-même avons fait les calculs les plus pessimistes en tablant sur une destruction des conteneurs renfermant une masse monolythique d'enrobés. En conclusion, je ne pense pas que les risques de fluage constituent un problème au niveau du stockage définitif.

W. HILD, R.F. d'Allemagne

A mon avis, M. Lefillatre a très bien résumé la situation dans ce domaine. On pourrait peut-être ajouter que l'on se trouve dans des conditions quelque peu favorables parce que l'on dispose d'un matériau qui a des caractéristiques visco-élastiques et par addition de sels solides ou d'autres produits, il est possible de modifier, dans une certaine mesure, la viscosité de façon à obtenir des produits plus stables du point de vue du stockage à des températures normales et du point de vue du fluage au cours du stockage intermédiaire. Mais on ne peut pas exclure tout à fait le fluage parce que la matière dont on a augmenté la viscosité reste encore visco-élastique et l'on doit supposer qu'au cours du stockage final, on aboutira à la situation décrite par M. Lefillatre. Vous avez pu constater, dans les calculs que M. Smailos a présentés ce matin, que l'on tient compte des possibilités de fluage et du remplissage total de la cavité par le produit.

H. ESCHRICH, Eurochemic

La question de savoir comment stocker les fûts de produit final, soit en position verticale, soit en position horizontale, est naturellement liée à la fluidité du produit à base de bitume. En ce qui concerne le stockage temporaire, il faut bien entendu tenir compte de la fluidité, que les fûts soient stockés en position verticale ou horizontale. La durée du stockage temporaire à prévoir dépend de la politique de gestion des déchets adoptée dans la région, le pays, voire même le continent, cependant à Eurochemic, il n'a pas été possible de définir clairement à l'heure actuelle la durée du stockage intermédiaire sur le site. Nous avons donc dû choisir un matériau durable pour nos fûts (et j'en arrive à la quatrième de nos cinq questions qui sont toutes liées à l'optimisation économique et technique de la bituminisation des déchets) ; étant donné les coûts, nous avons hésité à choisir l'un des meilleurs matériaux connus pour leur résistance à la corrosion vis-à-vis des milieux ambiants habituels tels que l'eau et l'air, ce matériau étant l'acier inoxydable. L'acier inoxydable est certes un très bon choix mais il est cher. Les fûts en acier doux protégés par une couche de peinture sont très certainement les conteneurs métalliques les meilleurs marchés.

Nous avons également envisagé l'utilisation de fûts en acier doux galvanisés qui pourraient dans certaines conditions de stockage avoir une meilleure résistance à la corrosion que des fûts en acier doux protégés par une couche de peinture.

Le matériau que nous avons finalement retenu, sur la base de divers essais de corrosion, est un acier chromé ; il s'agit d'un matériau de type "sandwich" qui consiste en une couche mince de type acier inoxydable sur ses deux faces avec au milieu une couche plus épaisse d'acier doux normal. Ses propriétés et son prix par rapport à ceux des autres matériaux que nous avions envisagés ont été présentés hier.

Il nous est apparu que ces fûts en acier chromé représentaient le meilleur compromis possible entre le prix et les qualités requises, même si le composant en acier doux était largement corrodé après une période de stockage prolongé, on peut s'attendre à ce que le fût conserve ses dimensions et que la couche d'acier chromé interne reste intacte, empêchant ainsi le produit bitumeux de couler à l'extérieur des conteneurs et permettant la récupération des déchets par une grue manipulée à distance ou un chariot élévateur sans contact direct avec le contenu radioactif.

En ce qui concerne la question du fluage, je souhaiterais rappeler que les fûts sont fermés à l'aide d'un couvercle de 30 cm de diamètre qui n'est pas, cependant, étanche aux gaz de façon à permettre la libération des gaz de radiolyse ; si les fûts étaient stockés en position horizontale, il existerait un risque d'obstruction des espaces vides par le bitume qui fluerait et par conséquent une augmentation de la pression à l'intérieur du fût, ce qui pourrait se traduire par un débordement du bitume à l'extérieur du fût.

En conclusion, si un stockage intermédiaire de durée inconnue doit être envisagé, il faut utiliser des conteneurs qui seront mécaniquement stables et résistants à la corrosion, stockés de préférence en position verticale.

N. VAN DE VOORDE, Belgique

Nos discussions portent sur la fluidité du bitume ; or, un certain nombre de pays utilisent, pour incorporer des déchets métalliques et d'autres déchets solides, du bitume pur. A l'heure actuelle, la question qui se pose vraiment est de savoir si cette façon de procéder est bonne car ce bitume est réellement fluide et le demeure ; on peut en effet le presser comme un tube de pâte dentifrice. Le bitume pur constitue-t-il un moyen acceptable d'incorporer des matières radioactives hétérogènes ?

Y. SOUSSELIER, France

Vous parlez, je pense, M. van de Voorde, de déchets métalliques ? C'est en effet un sujet qui peut fournir matière à discussion mais je voudrais d'abord demander au Dr. Malášek s'il désire ajouter quelque chose sur la question du fluage.

E. MALASEK, Tchécoslovaquie

Dans l'ensemble, nous avons adopté une méthode légèrement différente pour évacuer les produits bitumineux : en dessous d'une certaine radioactivité, soit jusqu'à un niveau de l'ordre de 0,1 Ci/k nous nous contentons de recourir à l'enfouissement définitif. Nous ne cherchons pas à stocker temporairement ces matières. Dans ces conditions, nous ne tenons compte que du bitume lui-même et non pas des propriétés des conteneurs. Cela revient à dire, par exemple, qu'en Tchécoslovaquie nous n'employons même pas de fûts métalliques mais simplement des fûts en papier qui sont utilisés dans l'industrie chimique, car ce matériau permet d'assurer le transport, dans des conditions sûres, des matières bitumineuses du lieu d'origine au lieu d'enfouissement. D'autres pays socialistes suivent la même politique ; en URSS, on travaille encore sur une plus grande échelle quelques expériences ont été consacrées au transport des produits bitumineux sous forme liquide dans la zone d'évacuation des déchets, où ils sont coulés dans des grandes cuves de stockage bétonnées ou directement dans le sol. Cependant, aucune décision finale n'a encore été prise dans le cas d'une activité plus élevée, soit de l'ordre de 1 à 10 Ci/kg de produit ; nous utilisons en général un conteneur métallique en acier doux et les produits sont évacués dans des tranchées en béton. Ce mode d'évacuation est également considéré comme un enfouissement définitif mais il reste, le cas échéant, la possibilité de récupérer les matières à un moment donné dans l'avenir. A l'heure actuelle, nous estimons que cela ne sera pas nécessaire car les zones d'évacuation des déchets sont sélectionnées du point de vue géologique de manière à présenter une certaine capacité d'absorption et à répondre à d'autres conditions ; elles devraient, en conséquence, présenter toute sécurité à l'avenir.

N. FERNANDEZ, France

J'ai une remarque à faire. Je crois que le problème du fluage est important dans deux cas particuliers : le problème du stockage temporaire qui impose une reprise - il faut donc que la durée de ce stockage temporaire soit fonction de la qualité du récipient qui contient le bitume ou inversement, c'est un premier point. Le deuxième point concerne le stockage définitif : il y a le problème des émetteurs alpha pour lequel personne ici ne peut garantir que l'enrobage dans le bitume sera suffisant (compte tenu des 24.000 ans de période radioactive du plutonium) et il y a le problème d'une reprise de ces déchets qui peut s'imposer aux générations futures. Le problème du fluage est donc à prendre en considération pour ce cas particulier des émetteurs alpha. Je pense que pour les autres émetteurs bêta-gamma, le problème du fluage n'a pas beaucoup d'importance.

H. CHRISTENSEN, Suède

Nous avons beaucoup entendu parler du bitumage des différents types de solutions de sels divers au cours de ce Séminaire ; il n'a par contre pas été question du bitumage des déchets les plus couramment produits, dans les réacteurs à eau bouillante ou dans les réacteurs à eau pressurisée, soit les différents types d'échangeurs d'ions. Un participant au Séminaire serait-il prêt à dire quelques mots à ce sujet ?

G. LEFILLATRE, France

Je pensais en avoir dit brièvement quelques mots dans ma communication en ce qui concernait notre expérience sur le conditionnement et la solidification des résines organiques. Nous avons une expérience en ce qui concerne ce problème, qui se pose aussi bien au niveau des BWR que des PWR. Il n'apparaît pas qu'il y ait de très grands problèmes de solidification de ce type de déchets, en utilisant des procédés continus, notamment dans le cas d'une technique que j'ai décrite sur le bitumage direct de ces résines en suspension dans l'eau déminéralisée sous forme très diluée (d'une dizaine de grammes par litre) ou d'une suspension très concentrée (de 200 à 300 grammes par litre à peu près). Je ne parle pas de procédés discontinus qui tous, que ce soit avec une machine ou avec une autre, ont des temps de contact très courts. Il faut éviter d'avoir des décompositions importantes de résines organiques qui sont des produits thermo-labiles et obtenir la garantie d'une dessication totale, soit pratiquement moins de 1 % d'eau dans l'enrobé.

Il s'agit d'un problème technologique parce qu'il y a une rétention de l'eau, de même d'ailleurs dans le cas des filtres qui constituent le même problème ; du fait de leur surface spécifique, ils retiennent l'eau d'une façon beaucoup plus forte que pour des sels solubles. On rencontre un peu le même problème dans le cas des boues coloïdales. A l'heure actuelle, la Centrale de Barsebäck conditionne des suspensions de résines depuis un an en actif. M. Edwall est encore plus à même que moi de répondre positivement à votre question.

Y. SOUSSELIER, France

Quelqu'un désire-t-il dire quelque chose sur la question posée par M. van de Voorde, à propos de l'enrobage direct de matériaux métalliques par du bitume pur ?

N. FERNANDEZ, France

Cette question, bien entendu, tout le monde se la pose ou se l'est posée. Il est très difficile d'enrober correctement des matériaux métalliques ou solides par du bitume puisqu'il peut y avoir de l'eau, et le bitume est coulé à chaud. Cela pose des problèmes de mise en oeuvre ; il peut y avoir formation de poches d'eau ; l'enrobage est difficile à contrôler. Il faudrait sécher auparavant et opérer à chaud, à la température de la coulée de bitume. De toute manière, l'utilisation du béton dans ce cas-là peut s'appliquer de façon très correcte. On peut faire du très bon béton armé avec de l'acier et c'est d'un coût relativement faible, ce qui rejoint d'ailleurs la question de l'économie globale et de l'optimisation de la gestion des déchets.

G. LEFILLATRE, France

En ce qui concerne ce problème de fluage, on se trouve confronté au problème du conteneur et je voudrais évoquer un double aspect de ce problème parce que l'on se heurte à deux conceptions différentes : l'une, qui est la nôtre et qui est celle d'Eurochemic, à savoir, d'avoir des stockages temporaires importants - puisque là on parle de cinquante ans pour Eurochemic et dix ans en ce qui nous concerne - et par ailleurs, la conception de certaines sociétés d'électricité - je pense notamment à l'Electricité de France - qui dès que leurs produits sont solidifiés, souhaitent s'en débarasser au plus vite, directement vers un stockage définitif. On se trouve à ce moment-là confronté à un problème de transport parce qu'on peut avoir des déchets qui peuvent atteindre 1 Ci/l et des blindages sont alors nécessaires ; qui dit blindage, dit château de plomb ou château avec du béton lourd - une infrastructure et des moyens de manutention importants pour pouvoir assurer le transport vers le site de stockage définitif. Il y a donc un choix à faire. Ce problème a d'ailleurs été rencontré à Saclay où l'on ne souhaite pas garder les déchets sur le site ; ils sont en fait transportés par châteaux blindés. On a adopté la solution de récupération de ces châteaux blindés et une rotation permanente, ce qui est à mon avis la bonne solution, car on garde l'un des intérêts premiers du bitume, c'est-à-dire la réduction de volume.

Il y a une autre possibilité qui consisterait à couler définitivement, à la sortie d'un poste d'enrobage quel qu'il soit, les déchets dans un conteneur en béton lourd qui aurait une protection variable en fonction de l'activité spécifique des déchets, et de considérer ensuite l'ensemble comme des déchets sous leur forme définitive. Si l'on adopte une telle solution, le problème du fluage n'existe plus.

N. FERNANDEZ, France

Je reviens toujours à la question du stockage définitif, et surtout à l'exemple qui a été présenté par l'Allemagne fédérale, c'est-à-dire le stockage en vrac dans une caverne de sel. Là se retrouve l'interaction entre les divers stades de la gestion des déchets. Si, dans ce cas particulier de l'Allemagne fédérale, les interactions sont favorables lorsque l'on ajoute le conditionnement par le bitume à la sécurité du stockage dans la mine de sel, ce n'est pas nécessairement vrai dans tous les cas. L'exemple du stockage en vrac n'est pas bon car il faut penser à la troisième barrière si la reprise s'impose et je pense toujours aux émetteurs alpha.

Y. SOUSSELIER, France

Je propose d'en terminer là sur cette question de fluage et de passer à ma deuxième question, sur le problème de la qualité du produit. A plusieurs reprises, on a parlé de l'intérêt ou de la possibilité éventuelle de séparer certains éléments de façon à améliorer la qualité du produit. On a envisagé aussi d'avoir ce que l'on peut appeler un double enrobage, puisqu'il peut y avoir un enrobage externe par du bitume pur. Cela pose d'une façon plus générale le problème de savoir jusqu'où doit-on aller dans la recherche de l'amélioration de la qualité d'un produit.

W. HILD, R.F. d'Allemagne

En ce qui concerne les caractéristiques des produits, je pense qu'elles sont, comme dans le premier cas que nous venons d'examiner, étroitement liées aux sites retenus pour l'évacuation définitive des matières. Dans ces conditions, toutes les qualités des produits, qu'il s'agisse de la résistance à la lixiviation, de la stabilité sous rayonnement ou de la stabilité thermique, doivent être considérées par rapport à la zone dans laquelle on se propose de les évacuer. A mon avis, nous avons acquis suffisamment d'expérience pour accepter même des matières qui, à première vue, ne répondent pas entièrement aux hypothèses que nous avons formulées en fonction du site d'évacuation. Comme il a déjà été indiqué, on peut agir sur les compositions chimiques ; c'est ainsi que, dans le cas d'Eurochemic, il est possible d'appliquer une procédure de traitement chimique pour aboutir à un produit qui soit à la fois plus résistant à la corrosion et plus résistant à la lixiviation. Reste le problème de l'enrobage qui est assuré, dans un cas, par l'enceinte de confinement et, dans l'autre, par une couche supplémentaire de bitume. Nous avons, à mon avis, suffisamment de moyens -

et je laisse à l'auditoire le soin d'en juger - pour agir et jouer sur la qualité conformément aux hypothèses admises. Je souligne à nouveau que nos hypothèses en matière de caractéristiques des produits sont principalement déterminées par les conditions propres à notre site d'évacuation. A cet égard, il convient aussi d'envisager les autres caractéristiques, soit la stabilité thermique des produits incorporés dans la matrice en bitume et la quantité de nucléides radioactifs, émetteurs alpha ou bêta-gamma, dans la mesure où elles sont liées à la stabilité sous rayonnement.

N. VAN DE VOORDE, Belgique

En entendant par exemple M. Lefillatre dire qu'il ne fallait pas décomposer les échangeurs d'ions et les conserver ainsi dans le bitume, de même que le nitrate de soude, les borates, etc., produits qui d'ailleurs ne sont pas radioactifs, je me suis demandé s'il ne fallait pas se pencher sur le problème des déchets de moyenne activité que nous avons tous contribué à créer. Est-ce qu'on ne pourrait pas s'attaquer au problème de la production de ce type de déchets ? Est-ce qu'il est absolument impossible d'éviter d'enrober les échangeurs d'ions dans le bitume ? Est-ce qu'on ne peut pas éviter d'enrober les sels solubles dans le bitume ? Il y a d'autres moyens possibles. Est-ce qu'on ne peut pas régénérer les échangeurs d'ions et précipiter les déchets à partir de ces solutions de régénération et n'avoir que des hautes activités à vitrifier directement, et la basse activité où tout le monde au fond est d'accord pour qu'on utilise le bitume ? Le problème du fluage ne se poserait plus tellement. Il y aurait simplement la basse activité pour laquelle l'enrobage dans le bitume est valable. Il y aurait la haute activité pour laquelle la vitrification s'imposerait. Entre les deux, c'est notre travail pour l'avenir.

E. ESCHRICH, Eurochemic

Nous avons déjà rejoint les rangs de ceux qui veulent s'attaquer à ce problème. Nous pouvons établir une distinction entre les déchets de faible activité et les déchets de moyenne activité, entre les déchets à vie courte et les déchets à vie longue et nous pouvons imaginer des combinaisons de ces diverses catégories. Quoique nous fassions, la politique de la gestion des déchets est en fin de compte un problème de stockage ou d'évacuation. Nous conditionnons les déchets dans le but d'assurer un confinement plus sûr sur le site de stockage de telle sorte que nous atteignons véritablement le but ultime de toute activité en matière de gestion des déchets qui est d'éviter que des doses d'exposition aux rayonnements atteignent l'homme ou d'autres êtres vivants à partir de sources radioactives externes ou internes. Ainsi que le relève M. Hild, il faut toujours en fin de compte examiner les relations entre les qualités du produit et les conditions du stockage.

S'il existait un lieu d'évacuation offrant une garantie quasi illimitée, nous n'aurions en aucune façon besoin d'envisager un conditionnement élaboré. Si nous avons des doutes en ce qui concerne la sûreté du conditionnement, des mesures appropriées doivent être prises pour contrebalancer tout facteur d'insécurité et si nous ne savons pas du tout où et quand les déchets seront finalement stockés ou évacués, nous devons appliquer les meilleures techniques de conditionnement des déchets qui sont disponibles à l'heure actuelle.

Ce qui fait habituellement l'objet de doute en ce qui concerne les considérations sur la sécurité à long terme est l'aptitude de l'homme à prévoir des situations sur terre dans des périodes éloignées de quelques centaines de milliers d'années et plus. Des modifications imprévisibles du climat, des sociétés, de la géologie, etc., peuvent remettre en question tout ce que nous avons fait dans l'intention d'isoler d'une façon sûre les déchets sur des périodes nécessaires.

Le fait d'examiner le risque à long terme conduit inévitablement au problème lié à la puissance des émetteurs alpha dans les divers types de déchets. Certains types de déchets de réacteurs sont exempts d'actinides et ne constituent pas un risque après la période de stockage relativement brève, tandis que les déchets qui résultent des opérations de retraitement sont toujours suspectés de contenir quelques émetteurs alpha.

Les déchets habituels de moyenne activité provenant du retraitement n'ont pas la même composition, en ce qui concerne les émetteurs alpha, que les solutions des hautes activités de produits de fission tout simplement parce qu'il provient d'autres étapes du procédé. Ici nous avons affaire principalement au plutonium, à l'uranium, ainsi qu'à quelques americium qui résultent de la décroissance du plutonium 241. Il est certainement possible, et nous l'avons prouvé par des études de laboratoire, mais pas encore à l'échelle industrielle, de séparer d'une façon quasi quantitative tous les actinides des solutions des déchets de moyenne activité en faisant appel à des méthodes chromatographiques. Nos essais ont indiqué, en outre, que la séparation des produits de fission des quantités importantes de sels inertes, généralement du nitrate de sodium, est réalisable par des moyens relativement simples. Cependant, la séparation des émetteurs alpha au niveau requis se traduirait déjà par un produit bitumeux amélioré qui nécessiterait seulement une période relativement courte de stockage - tout au moins une période pendant laquelle il est possible d'avoir un contrôle avec une bonne certitude jusqu'à ce que les produits de fission contenus aient décru à des niveaux inoffensifs.

W. HILD, R.F. d'Allemagne

Je souhaiterais simplement déclarer, en complément aux remarques de M. van de Voorde et de M. Eschrich, que je suis très partisan de ramener les problèmes de déchets auxquels nous sommes confrontés à deux catégories de problèmes posés respectivement par les déchets de haute activité et par les déchets de faible activité. Nombreux sont mes anciens collègues d'Eurochemic présents à cette réunion et ils savent bien que nous avons longuement débattu cette question. En fait, nous avons aussi, à Karlsruhe, finalement entrepris un programme qui vise à orienter les activités dans ces deux directions distinctes, soit, les déchets de haute activité et les déchets de faible activité. Il serait très intéressant et très judicieux, à l'avenir, de répartir les déchets de moyenne activité entre ces deux catégories. Cependant, il ne faudrait pas cesser de souligner que la gestion des déchets devrait commencer à s'appliquer à l'endroit même où les déchets sont produits. Dans toutes les installations nucléaires, les responsables tant de la conception que de l'exploitation devraient se rendre compte qu'ils produisent des déchets et ils devraient faire de leur mieux pour concevoir les procédés de façon à ce que la production de déchets soit maintenue à un volume aussi faible que possible et pour les exploiter de manière à ce que la plus grande quantité possible de matières puisse être recyclée. Si les spécialistes de la gestion des déchets et tous ceux qui font fonctionner et exploitent des réacteurs de puissance et des installations de retraitement unissent ainsi leurs efforts, je pense que nous aurons à l'avenir une bonne probabilité de trouver des procédés nouveaux, voire plus efficaces, pour le traitement et l'évacuation des déchets.

E. MALASEK, Tchécoslovaquie

On a déjà beaucoup parlé de l'influence des conditions d'évacuation sur la production de matières bitumineuses. Cependant, un autre facteur n'a pas encore été abordé : il s'agit de notre tendance à incorporer un nombre aussi élevé que possible de concentrats radioactifs dans une quantité minimale du produit final. Ce facteur à nouveau peut influer sur les conditions d'évacuation. On observe parfois un facteur supplémentaire qui est lié aux conditions imposées par le procédé lui-même ; c'est ainsi que, dans certaines associations de sels ou de boues, il est possible d'avoir des concentrations beaucoup plus importantes du point de vue du stockage mais la température ou le procédé retenu pour le traitement des matières, voire le bitume lui-même, ne se prêtent pas à l'incorporation de plus grandes quantités de ces matières. Un autre moyen d'améliorer les propriétés des produits consiste à réduire au minimum la quantité de matières solubles qui sont comprises dans les produits bitumés. Je souscris à la question posée par M. van de Voorde au sujet du bitumage des résines échangeuses d'ions. Nous avons beaucoup travaillé dans ce domaine et j'aurais aussi tendance à dire qu'il est possible d'incorporer des résines échangeuses d'ions dans du bitume mais nous ne sommes pas encore assurés des propriétés de ces produits, notamment pour le stockage à long terme. A l'heure actuelle, nous ne sommes pas réellement en mesure de donner une réponse définitive et nous avons tendance à réduire au minimum le volume de résines échangeuses d'ions qu'il est nécessaire d'évacuer. Personnellement, je préférerais la méthode qui a été exposée à Vienne par M. van de Voorde au sujet de l'incinération de ces matières avant bitumage.

G. LEFILLATRE, France

Je voudrais faire trois commentaires. Le premier porte sur la qualité du produit fini en fonction du stockage. Il est évident qu'il faudrait, avant de définir un procédé de solidification, savoir quel est le stockage définitif qui sera assuré pour ce type de déchets. Mais il n'empêche que quel que soit le mode de stockage définitif et le lieu où va être situé ce stockage, il y aura toujours un problème de transport. Il me semble exclu de ne pas envisager une solidification des déchets, ne serait-ce que pour assurer d'une façon sûre le transport entre le lieu de production et le lieu de stockage définitif. Par conséquent, le bitumage peut, si c'est un procédé économique, remplir ces conditions au même titre que d'autres procédés, puisqu'encore une fois il présente l'avantage de la réduction de volume. Dans ce même ordre d'idées, s'il s'agit d'un stockage très sûr comme les mines de sel, on ne voit pas l'intérêt d'étudier une amélioration possible de ces déchets puisque l'on dispose déjà d'une très grande sécurité du point de vue du stockage définitif. Il ne reste plus que la sécurité en matière de transport.

Ma deuxième remarque concerne les résines et je suis tout à fait d'accord avec le commentaire de M. van de Voorde ; malheureusement, je constate que ce problème des résines usées est vraiment typique des réacteurs à eau légère. Dans les usines de retraitement, nous avons des résines échangeuses d'ions qui traitent des activités beaucoup plus élevées que dans les réacteurs, ou tout au moins du même ordre de grandeur dans certains domaines, et jusqu'à présent nous n'avons jamais rencontré de problèmes avec des résines usées parce qu'on les régénère systématiquement (je parle dans le cadre de la France tant à Marcoule qu'à La Hague). Je crois que ce problème est à voir avec les compagnies d'électricité, car c'est peut-être aussi un problème économique, étant donné que les résines échangeuses d'ions coûtent de plus en plus cher et que le volume de résines qui sont perdues est très élevé.

Enfin, troisième remarque, toujours sur le commentaire de M. van de Voorde sur le fait d'enrober dans le bitume des sels qui n'ont rien de radioactif. Je voudrais faire remarquer que son commentaire ne vas pas tellement dans le sens actuel, à savoir un "rejet zéro", parce qu'on va à ce moment-là être amené à faire une chimie très compliquée, et que par ailleurs on ne sera jamais certain d'avoir des sels qui seront totalement inactifs. Il se posera donc toujours le problème du conditionnement ultime de ces sels. Bien entendu, une séparation notamment pour les émetteurs alpha serait souhaitable, de façon à conditionner sous un volume réduit le maximum d'activité et une telle mesure serait positive. Enfin, en supposant même que l'on ait des facteurs de décontamination très élevés, on aura toujours un déchet de faible ou de très faible activité qu'il ne sera pas possible de rejeter dans le milieu sans conditionnement.

N. FERNANDEZ, France

Je pense que la qualité du procédé d'enrobage est importante parce qu'elle permet de résoudre certains problèmes difficiles comme le confinement des résines échangeuses d'ions, le confinement de solvents lourds comme la TLA, etc. S'il faut poursuivre l'étude de la qualité de ces enrobés, c'est pour résoudre les problèmes qui peuvent se poser et qu'il faut toujours prendre en considération parce qu'il y aura toujours des fonctionnements anormaux. Je sais bien que M. van de Voorde, M. Hild et M. Malášek ont raison sur le fond, mais il y aura toujours des fonctionnements anormaux, des problèmes à résoudre dans ce domaine de la moyenne activité, car bien souvent on ne dispose pas d'un système global de traitement qui permette de passer d'une catégorie d'activité à l'autre. Ensuite, il y a un aspect de la question qualité qui n'a peut-être pas été posé : c'est celui du "suivi" de la qualité du produit élaboré par les installations. Je ne pense pas qu'il faille faire ce "suivi" en fonction de la qualité optimale qu'on peut obtenir avec le procédé, mais plutôt en fonction de la qualité moyenne qui est suffisante pour assurer la sécurité dans les stockages futurs. Il faut que l'ensemble conduise à un "suivi" relativement peu onéreux. S'il faut faire partout des installations de contrôle aussi poussées et élaborées que celles d'Eurochemic, cela revient horriblement cher. On rejoint là l'aspect économique et il faudrait faire un "suivi" relativement peu coûteux, pour obtenir une qualité moyenne suffisante

J. ORTEGA ABELLAN, Espagne

Je crois qu'il y a un problème avec la conversion des nitrates en nitrites. Les nitrates avec l'hydrogène produit par radiolyse peuvent se transformer en nitrites qui constituent un problème pour la sécurité. Je crois que dans le temps, les quantités de nitrites peuvent devenir importantes.

G. LEFILLATRE, France

En ce qui concerne le problème de la formation des nitrites dans les solutions de produits de fission, les américains ont déjà posé ce problème il y a un certain nombre d'années, autour des années 65-66, après le développement des centres de retraitement et l'augmentation des puissances des réacteurs.

D'après les études faites à Oak Ridge, il s'était avéré qu'il y avait des problèmes très graves en ce qui concernait les mélanges de bitume et de nitrite de sodium dans certaines proportions. Pour notre compte, nous avons refait des essais systématiques avec des mélanges très variables, en mettant aussi des quantités de bitume très faibles, de façon à tenir compte des problèmes d'encroûtage dans les appareils ; par exemple à hauteur de 90 % de sel, ce qui est assez utopique, et 10 % de bitume, on n'a jamais observé de problèmes particuliers d'exothermie. Les problèmes que j'ai cité dans ma communication étaient dûs surtout au nitrate d'ammonium et je ne pense pas personnellement qu'il y ait un problème avec le nitrite de sodium.

W. HILD, R.F. d'Allemagne

Je souhaiterais souligner l'argument que M. Lefillatre vient d'invoquer. Notre expérience est plus ou moins la même que la sienne et j'estime que le problème des nitrites ne doit être considéré comme grave que s'il s'agit de déchets qui sont réellement de haute activité, de sorte que de fortes doses de rayonnement entraînent, par destruction radiolytique, la formation de nitrites. Le danger, comme nous le savons tous, réside dans un système où se trouve le nitrite d'ammonium car celui-ci peut réellement se décomposer facilement et jouer même le rôle d'explosif. En conséquence, ce système doit être évité mais il existe suffisamment de moyens d'éliminer l'ammonium. Hormis ce phénomène, en ce qui concerne les déchets de moyenne et de faible activité, je ne pense pas qu'il s'agisse là d'un problème important.

H. ESCHRICH, Eurochemic

Je souhaiterais simplement ajouter que la quantité ou la concentration de nitrites que l'on pourrait craindre de trouver dans la solution constituant en fin de compte le produit à base de bitume, dépend dans une très large mesure de la façon de stocker les déchets liquides. Si ces derniers sont stockés en milieu acide, l'acide nitreux présente alors une stabilité et une solubilité limitées ; je ne connais pas exactement par coeur les concentrations à l'équilibre. Elles dépendent naturellement de l'acidité libre totale et de la température, mais elles ne dépassent pas quelques centièmes de mole en solutions aqueuses. Si les solutions sont stockées en milieu alcalin, les quantités de nitrites risquent de s'accroître. D'autre part, si l'on craint que la présence de nitrites puisse créer des problèmes, ce qui est très rare ainsi que nous venons de l'entendre, il est très facile de les détruire soit par simple barbotage des solutions acides ou acidifiées, soit à l'aide de réactifs qui réagissent quantitativement et rapidement avec le nitrite et conduisent à des produits qui sont complètement inoffensifs tels que l'azote et l'eau par exemple. Je ne pense pas, par conséquent, que les nitrites constituent un problème susceptible de soulever des difficultés notables. Celles ci peuvent être aisément surmontées si l'on craint que certaines solutions contenant des nitrites puissent constituer un risque au cours d'enrobage dans le bitume.

R. SIMON, C.C.E.

M. Lefillatre pourrait-il éventuellement chiffrer les coûts de l'enrobage des résidus des centrales à eau bouillante, par exemple, des résines amberlite ou des résines en microsphères ?

G. LEFILLATRE, France

Il est difficile de donner un prix "ex abrupto" étant donné qu'il y a énormément de conditions qui jouent. Nous avons, au CEA, des prix de l'ordre de 900 FF/m^3, y compris le transport, et la collecte des effluents dans le Centre, la concentration par évaporation, le conditionnement par le bitume et le transport ultérieur au site de stockage. Disons qu'en ordre de grandeur ce coût est pour Marcoule d'environ 10.000 FF/m^3 de déchets bien définis.

Ceci est voisin des chiffres donnés par Eurochemic hier qui étaient de l'ordre de 12.000 FF/m^3.

Y. SOUSSELIER, France

S'il n'y a pas d'autres questions sur le sujet de la qualité des produits, je voudrais que nous passions à la troisième question que je vous ai proposée : le problème du risque maximum, le problème de l'incendie. On nous a montré ce matin qu'il ne faut peut-être pas dramatiser la question ; il n'en reste pas moins vrai que le bitume peut effectivement brûler et que ce problème peut se poser durant les opérations de fabrication, ce qui semble assez facile à maîtriser; pendant les transports où là aussi il est peut-être assez facile à maîtriser; mais surtout pendant le stockage temporaire et c'est cela que je souhaiterais que nous discutions.

Doit-on dans le cas de stockages temporaires prévus pour durer plusieurs dizaines d'années, comme pour Eurochemic, prendre en compte le risque d'incendie et peut-on agir sur la qualité du produit, même si on a un produit qui coûte plus cher ? Peut-on faire en sorte d'éviter de prendre des mesures, comme celle qui consiste à avoir des bunkers de stockage extrêmement élaborés contre la chute d'avions ?

H. ESCHRICH, Eurochemic

Le fait que les déchets radioactifs incorporés dans le bitume puissent brûler a toujours été considéré comme l'un des inconvénients majeurs du bitumage, et c'est la raison principale pour laquelle certaines personnes préfèrent utiliser d'autres procédés de traitement. Les très rares incidents qui ont entraîné la combustion des produits à base de bitume sont, jusqu'à présent, toujours survenus pendant le procédé d'incorporation lui-même, ou juste après. Nous savons maintenant qu'il existe suffisamment de moyens de remédier au risque d'incendie aux points critiques des installations. Cependant, le risque d'incendie pendant le stockage intermédiaire demeure, et diverses mesures peuvent être prises pour limiter ce risque à un minimum acceptable. L'une d'entre elles pourrait consister à procéder à une sorte d'enfouissement dans une enceinte artificielle de stockage par exemple en remplissant les espaces libres entre les fûts ou les autres emballages par du sable ou une autre matière non-combustible qui, lorsqu'il serait nécessaire de récupérer les déchets, pourrait être facilement retiré par aspiration, notamment si l'on est certain qu'il n'existe pas de radionucléides ou de matériaux non fixés dans cette matière destinée à protéger contre les risques d'incendie. Peut-être cette mesure est-elle un peu trop sophistiquée. Notre point de vue à Eurochemic est le suivant : en cas d'incendie, il y a tout d'abord une chose qui peut brûler et, ensuite, une autre qui la fait brûler. La chose qui brûle est le déchet et nous savons éliminer ce qui le fait brûler. Ceci est le principe que nous allons adopter pour le stockage intermédiaire dans nos enceintes : il ne restera dans ces enceintes aucune installation ou aucun équipement qui, une fois les enceintes remplies, puisse être une source d'étincelles ou de flammes. Nous n'avons pas d'éclairage, pas d'installation ni d'appareil électrique, rien qui puisse produire des étincelles, ni aucun élément mobile. L'électricité statique est normalement évitée dans une atmosphère qui est totalement ionisée, ce qui est certainement le cas dans le champ de rayonnement qui existe dans les enceintes de stockage. Une ventilation forcée évite toute accumulation de mélanges de gaz explosifs. Nous ne croyons pas, par conséquent, qu'il y ait des risques d'incendies dans la mesure où le système d'ensemble du stockage à l'intérieur du confinement est en cause. Des précautions doivent être prises pour permettre le refroidissement des surfaces des enceintes par de l'eau à partir de l'extérieur et pour combattre un incendie qui se propagerait à l'extérieur des enceintes de stockage. Si les déchets bitumeux brûlent, il y a généralement une phase gazeuse ainsi que des cendres résiduelles. La combustion de produits bitumeux a fait l'objet de recherches ainsi que de certaines expériences à GfK à Karlsruhe: de l'ordre de 20 à 30 % des mélanges des produits de fission incorporés étaient "volatilisés", tandis que le reste des matières était retrouvé dans les cendres résiduelles. Par conséquent, même un incendie, notamment dans une enceinte de confinement qui a été dimensionnée et construite compte tenu de ce risque, n'entraînera pas nécessairement la dispersion de radioactivité dans l'environnement; mais je répète que dans le cas de notre stockage temporaire, nous avons pris principalement des mesures pour éliminer les causes d'une combustion des produits stockés plutôt que d'essayer de combattre l'incendie.

W. HILD, R.F. d'Allemagne

Je souhaiterais déclarer une fois de plus que nous sommes tous conscients du fait non seulement que le bitume et les produits à base de bitume sont des matières combustibles mais aussi que ces matières ne sont pas facilement inflammables et qu'il existe des moyens suffisants de les protéger, même pendant le stockage temporaire, pour éviter qu'elles ne prennent feu. Comme je vous l'ai montré au cours de mon exposé, il faut une source extrêmement chaude pour faire brûler un bloc de bitume. C'est pourquoi nous sommes assez confiants, compte tenu notamment des possibilités qu'il y a de combattre un incendie et de contrôler les zones de stockage.

Y. SOUSSELIER, France

Je voudrais demander l'avis de M. Fernandez sur ce point puisqu'il est directement impliqué dans ces problèmes de stockage.

N. FERNANDEZ, France

Je pense, de même que M. Hild, qu'il est très difficile de faire brûler du bitume. Il faut apporter une source d'énergie très importante. Il s'agit vraiment d'un accident de dimensionnement, c'est-à-dire la chute d'un avion rempli de kérozène qui fait brûler le stockage en apportant une source de chaleur considérable. La probabilité en est sans doute très faible, mais il faut quand même le considérer.

Y. SOUSSELIER, France

Cela veut-il dire que vous pensez qu'il faut le prendre en considération ?

N. FERNANDEZ, France

Si les Commissions de sûreté locales d'un pays disent qu'il faut prendre en considération des accidents de dimensionnement, il est nécessaire de les prendre en considération.

R. SIMON, C.C.E.

M. Eschrich a indiqué qu'en cas d'incendie une certaine fraction de produits de fission s'échappera sous forme de matières volatiles. Qu'arrivera-t-il à cette fraction volatile ?

W. HILD, R.F. d'Allemagne

M. Eschrich a évoqué des expériences que nous avons effectuées en collaboration avec l'Institut pour la chimie des propergols et des explosifs. Une partie des diapositives projetées au cours de mon exposé ont été prises au cours de ces recherches. Elles montraient que nous avons expérimenté diverses mesures de lutte contre l'incendie. Nous souhaitions également nous faire une idée de la dispersion de la radioactivité dans le cas d'un produit n'ayant pas d'action sur l'incendie. Sans lutter contre l'incendie, nous avons fait brûler jusqu'à la fin un produit ayant une forte teneur en nitrate (fût de 175 litres rempli, à raison de 50 % environ de son poids, d'un mélange de nitrate de sodium et de bitume). Pour nous faire une idée de la dispersion dans les environs, nous avons recueilli les matières retombant avec les fumées et mesuré la vitesse et la direction du vent; les valeurs citées par M. Eschrich sont d'un ordre de grandeur exact. Nous avons analysé les matières contenues dans les cendres du fût et le reste du fût, les avons dénombrées et établi un bilan des matières découvertes et recueillies dans les différents récipients se trouvant aux alentours. Comme la teneur en nitrate de sodium s'est révélée être très élevée, nous estimons (comme suite à l'analyse effectuée) que ce phénomène de volatilisation n'est pas de la volatilisation à proprement parler; il s'agit plus ou moins d'un transfert de poussières d'hydroxyde de sodium sous forme d'aérosols. En fait, ce transfert est de l'ordre de 20 % mais, si la question vous intéresse, je peux vous donner les chiffres exacts.

K.A. TALLBERG, Norvège

Je souhaiterais savoir, à propos de cette expérience d'incendie, si vous avez continué à ajouter de l'essence afin que la combustion se poursuive ou si le produit a brûlé tout seul pendant la durée de l'expérience ?

W. HILD, R.F. d'Allemagne

Pour cette expérience particulière, nous avons utilisé les connaissances acquises lors de la première expérience de combustion. Nous avons ajouté une quantité suffisante de pétrole pour déclencher l'incendie et l'amener à un point où il s'alimentait de lui-même.

Y. SOUSSELIER, France

Nous pourrions peut-être passer maintenant au point suivant qui est une question que nous avons déjà un peu effleurée: le problème de l'optimisation globale du procédé, et la question de savoir si nous sommes arrivés à présent au point où l'ensemble de la gestion et la façon dont est réalisé le bitumage sont optimisés, étant entendu que l'on peut concevoir cette gestion comme allant jusqu'au stockage final. Si on ne peut pas prendre en compte le stockage final, on peut considérer que le stockage temporaire est compris.

Ma question comporte en particulier les problèmes liés à la séparation de certains produits, ce qui par exemple entraînera des frais supplémentaires mais se traduira peut-être par des coûts moindres ensuite, du point de vue du stockage ou des exigences des autorités de sûreté. Ma question est assez délicate; elle nous amène aux questions de coûts qu'il est toujours assez difficile d'estimer, mais je pense qu'il serait quand même intéressant que nous en disions un mot.

G. LEFILLATRE, France

Je pense que le problème principal sur l'avenir des études relatives au bitumage des déchets radioactifs tient essentiellement à la question suivante: jusqu'à quelle limite peut-on raisonnablement envisager de bitumer les déchets radioactifs ? Nous avons vu qu'actuellement les différents orateurs parlaient de 1 Ci/l comme étant une limite raisonnable actuelle. Il y a quelques années, on parlait de 10 Ci/l. A mon avis, personne n'en sait rigoureusement rien. C'est donc le problème principal qui reste à étudier.

Le deuxième problème, ce sont les émetteurs alpha: jusqu'à quelle limite également pourrait-on conditionner par le bitume ce type de déchets de façon à avoir de bonnes garanties pour le stockage à long terme ?

Voilà les deux grands axes d'études à mener dans un proche avenir pour le conditionnement dans le bitume des déchets radioactifs.

Y. SOUSSELIER, France

Je vais quand même vous reposer une question: comment pensez-vous que l'on puisse orienter ces recherches, car ce sont des sujets de recherche très généraux ? Quel programme à votre avis pourrait-on mener, par exemple pour déterminer les limites jusqu'où on peut aller, jusqu'à 1 Ci/l ou jusqu'à 10 Ci/l, étant entendu que les études doivent être quand même menées assez rapidement compte tenu du développement de l'énergie nucléaire, et que dans certains cas, c'est le problème du comportement à long terme qui est posé ?

G. LEFILLATRE, France

Le programme français est actuellement un peut ralenti pour des raisons de délais et de coûts, mais nous espérons l'an prochain être à même de reprendre une expérimentation qui a été arrêtée en 1971, à savoir l'enrobage de produits de fission au niveau de 100 Ci/l, étant bien entendu qu'on n'envisage pas d'enrober dans l'avenir 100 Ci/l, mais d'avoir au moins un facteur 10 d'augmentation pour accélérer le processus d'auto-irradiation et pouvoir chiffrer, d'une façon plus valable à mon avis qu'actuellement, le dégagement des gaz de radiolyse et les problèmes d'échauffement - ceci pour la partie gamma surtout. Pour la partie alpha, nous avons aussi un programme de séparation des actinides, soit par des méthodes d'échanges d'ions, soit par des méthodes chimiques. Ensuite, à partir de cette séparation, nous pensons essayer, sous réserve de ne pas avoir de problèmes de criticité, d'enrober, un peu comme le font nos collègues de la vitrification, le maximum d'activité spécifique avec des émetteurs alpha (plutonium, curium, americium).

W. HILD, R.F. d'Allemagne

A mon avis, nous devrions poursuivre nos recherches relatives à la stabilité sous rayonnement. M. Lefillatre a fait remarquer qu'il serait intéressant de poursuivre ces expériences et nos collègues des pays d'Europe de l'Est ont entrepris quelques recherches dans ce domaine. Certains autres points restent à élucider en ce qui concerne la comparaison entre les essais d'irradiation externe et des essais d'irradiation interne. A cet égard, il serait du plus haut intérêt - et je pense que les expériences qu'il est prévu d'entreprendre à Eurochemic avec nos collaborateurs de Karlsruhe contribueraient notablement à la solution de ce problème - de chercher à étudier la stabilité des produits à base de bitume en liaison avec de plus fortes concentrations d'émetteurs alpha types. Nous pourrions peut-être prévoir une expérience analogue à celle qui est actuellement effectuée sur les produits vitrifiés; ce type d'expérience porte sur un certain laps de temps et il consiste à y incorporer du curium 242 dans le produit à base de bitume, afin de provoquer, dans un délai relativement bref, le dommage que pourrait subir, du fait du rayonnement, un produit à base de bitume qui serait stocké pendant une période correspondant à la désintégration complète des actinides présents. Ces questions méritent d'être étudiées.

H. ESCHRICH, Eurochemic

Ma contribution sera très théorique car nous ne savons pas exactement quels seront les frais de fonctionnement de nos installations et nous n'avons pas jusqu'à présent d'expérience pratique en ce qui concerne le fonctionnement en actif. Cependant, il n'est pas toujours nécessaire de disposer d'une installation analogue à celle d'Eurochemic pour traiter des déchets radioactifs. Je pourrais notamment imaginer sur le site d'un réacteur une installation de bitumage de configuration beaucoup plus simple et dont les frais d'investissement seraient, selon toute vraisemblance, moins élevés. En outre, l'importance des effectifs requis pour faire fonctionner cette installation correctement, dépend du nombre de dispositifs de contrôle et du degré d'automatisme dont on souhaite doter l'installation. De plus, la qualité du produit peut également être influencée par le montant des fonds disponibles. C'est ainsi que, comme nous l'avons déjà signalé, on peut aussi utiliser des réactifs meilleur marché, susceptibles de diminuer légèrement la qualité du produit qui resterait malgré tout à un niveau acceptable. Il n'est pas nécessaire de choisir des solutions extrêmes pour obtenir le meilleur produit possible. Si l'on souhaite procéder ainsi, quelques mesures pourraient être prises, par exemple pour lutter contre la combustibilité ainsi que l'on a essayé de le faire il y a quelques années aux Etats-Unis ; il existe des additifs qui permettent de diminuer l'inflammabilité et de retarder la combustion ainsi que le taux de combustion. En ce qui concerne l'activité spécifique maximale admissible, les discussions se poursuivent et nous allons fabriquer et essayer quelques produits d'activité spécifique élevée dans des dimensions industrielles de telle façon que nous puissions contribuer par les résultats obtenus à élucider ce problème sur une base véritablement industrielle.

Du point de vue théorique, si un type de bitume est stable sous rayonnement, à 10^9 rads par exemple, il est possible de calculer la dose cumulée que recevra une matrice en bitume de 1 Ci/l contenant des produits de fission mixtes et de constater que la dose cumulée sous l'effet d'une désintégration complète n'atteint pas 10^9 rads. Sur des bases théoriques, cela équivaudrait à 100 curies par litre environ ou à une dose quelque peu inférieure. Cependant, il s'agit là d'une théorie qui demande à être confirmée par des expériences.

E. MALASEK, Tchécoslovaquie

Je ne suis pas en mesure de donner des valeurs absolues concernant les procédés de bitumage. Nous avons effectué des travaux à ce sujet en évaluant l'économie du système de gestion des déchets dans son intégralité, lorsque nous sommes passés de l'ancien procédé d'enrobage dans le ciment à un procédé de bitumage et de l'ancienne à la nouvelle zone d'évacuation. Il s'en est suivi une diminution de 60 à 80 % des coûts totaux par rapport au procédé d'enrobage dans le ciment mais celle-ci était principalement imputable, en l'occurence, au coût du mètre cube de zone de stockage et au remplacement des conteneurs utilisés pour le ciment par des conteneurs destinés à recevoir du bitume. De façon générale, il s'est avéré que, par rapport à l'enrobage dans le ciment, le procédé de bitumage est plus économique. En ce qui concerne l'activité maximale susceptible d'être incorporée dans du bitume, il apparaît à l'heure actuelle que, jusqu'au niveau d'un curie par litre, tout bitume sera approprié et n'impliquera pas de prescriptions spéciales visant à éliminer les produits gazeux. Dans le cas des activités plus élevées, nous stockons des matières dont l'activité spécifique est de l'ordre de 10 à 50 curies par kg de produit. Dans ces cas, on utilise du bitume de type dur, la tendance étant d'obtenir un produit poreux qui ne se dilate pas et qui permette le dégagement des produits gazeux, lesquels sont éliminés de la zone de stockage au cours d'une phase initiale. On estime qu'après un certain temps, suivant l'activité, il serait possible d'arrêter le système de ventilation et de stocker les produits sans ventilation.

N. FERNANDEZ, France

En tant qu'exploitant, je me sens concerné. L'économie optimisée dans la gestion des déchets est toujours liée au même problème qui est celui du risque qui caractérise les déchets en fonction du procédé qu'on utilise. En ce qui nous concerne, c'est le bitume avec ses problèmes particuliers. Mais ce problème, qui est général, peut se décomposer en plusieurs sous-problèmes, notamment le problème du génie chimique des installations et je pense aux interventions et à l'entretien. Les installations devraient être conçues pour pouvoir être entretenues facilement et pour produire le moins possible de déchets secondaires. Il y a aussi le problème du "suivi" de la qualité du produit que l'on fabrique, et je crois qu'il faut se limiter au "suivi" nécessaire et suffisant pour garantir la qualité moyenne du produit obtenu. Il y a aussi le problème du risque potentiel, et je pense toujours aux émetteurs alpha il faut, à mon avis, faire supporter au producteur le complément du coût lié au type de stockage définitif nécessaire. Ceci pourra inciter le producteur à ne pas produire ce type de déchets. Et puis ensuite, il y a un problème de discipline générale, non seulement de la part du producteur, mais aussi des responsables à haut niveau qui doivent imposer cette discipline, et d'autres la respecter: c'est un problème de responsabilité. Si on pouvait atteindre ces objectifs, bien entendu on arriverait à optimiser cette économie; de toute manière la gestion des déchets coûtera toujours relativement cher.

Y. SOUSSELIER, France

Tout le monde sera volontiers d'accord avec ce que vous venez de dire. D'une façon plus générale, il me semble qu'au cours des dernières années, on s'est aperçu que la gestion des déchets coûtait plus cher qu'on ne l'avait imaginé au départ. Peut-être les augmentations dans ce domaine-là ont été moindres que dans d'autres domaines de l'énergie nucléaire, mais elles n'ont toutefois pas été négligeables.

N. FERNANDEZ, France

Je suis un peu inquiet, parce que plus on parle de procédés plus on cherche des procédés très élaborés, sophistiqués qui coûteront de plus en plus cher.

P.W. KNUTTI, Suisse

M. Eschrich a signalé l'existence de certains additifs permettant de réduire la combustibilité du bitume. Je me demande si cette technique pourrait être appliquée à l'échelle industrielle et de quels additifs il s'agit. M. Eschrich pourrait-il exposer cette possibilité plus en détail car, si elle existe, ce serait là un excellent moyen de traiter nos déchets.

H. ESCHRICH, Eurochemic

Ces additifs sont appelés quelquefois "remplisseurs". L'introduction de ces additifs ne signifie pas que le bitume devient incombustible, il demeure un produit combustible mais, ce que l'on a obtenu est un retard dans le démarrage de l'incendie. Je me réfère en l'occurence aux inhibiteurs d'incendie, comme l'hydroxyde d'aluminium ou les phosphates, que nous avons utilisés dans l'un de nos diagrammes de prétraitement. Plus l'on utilise de ces additifs, moins le bitume devient inflammable. On a alors besoin d'une flamme qui se maintient pendant un long moment pour déclencher la combustion Bien entendu, en réduisant la quantité de produits contenant de l'oxygène, on obtient un produit qui a une meilleure qualité que celle observée en général avec les déchets de moyenne activité provenant des installations de retraitement, par exemple. Nous devons procéder à une analyse approfondie des coûts et avantages pour décider s'il convient d'incorporer dans le produit à base de bitume des substances qui retardent l'incendie. Cependant, il faudrait bien se rendre compte que le conditionnement des déchets par une technique de bitumage y compris la manutention et le stockage temporaire, ne représente que quelques dixièmes du coût total de la gestion des déchets, la plus grande partie de ces coûts étant normalement imputable au transport et au stockage définitif. Cette proportion est théorique car personne ne connaît ces coûts, mais elle correspond approximativement à ce que l'on peut prévoir. En conséquence, il est possible d'utiliser des additifs appropriés pour améliorer le produit dans les cas où l'on ne fait guère confiance aux conditions de sécurité des sites de stockage temporaire ou définitif qui ont été retenus.

P.W. KNUTTI, Suisse

De quelle façon ce procédé influe-t-il sur les autres propriétés des matières, telles que le taux de lixiviation, et a-t-il un effet négatif ?

H. ESCHRICH, Eurochemic

Si des additifs solides destinés à réduire les risques d'incendie sont ajoutés, il en résultera un accroissement de la teneur relative en solides dans le produit final ou bien une augmentation en volume de celui-ci. Si un pourcentage élevé de ces produits doit être ajouté pour obtenir l'effet désiré, on doit s'attendre à ce que le taux de lixiviation de certains nucléides soit influencé d'une façon négative, étant donné que la couche disponible pour l'enrobage des solides par le bitume devient plus faible.

Cependant, on peut également ajouter des matériaux qui diminuent la combustibilité en même temps que la lixiviation des produits et qui agissent en outre comme des agents de nettoiement ou des sorbants. Le fait que ces additifs puissent affecter d'une façon négative ou positive les propriétés du produit final dépend de leur nature, des quantités ajoutées ainsi que de leur répartition à l'intérieur de la matrice de bitume. S'ils sont choisis avec soin, ils ne diminueront pas les propriétés des produits qui sont essentielles pour un stockage sûr à long terme.

Enfin, nous devons toujours prendre en considération la combinaison de la qualité des produits bitumeux et de la qualité du stockage de façon à obtenir le niveau de sécurité requis.

Y. SOUSSELIER, France

C'est toujours un problème d'optimisation ! D'autres questions ?

E. BREGULLA, R.F. d'Allemagne

Je souhaiterais formuler deux remarques au sujet de l'économie. En premier lieu, si l'on regarde ce qui se passe dans différents laboratoires et dans d'autres établissements, il apparaît que l'utilisation de grandes quantités d'eau en vue de réduire la radioactivité présente beaucoup d'analogies avec l'utilisation de sels et autres détergents. En second lieu, nous devons nous préoccuper davantage de la conception et de la construction du matériel destiné à l'entretien et au remplacement car, en économisant de l'argent et du temps au cours de cette phase, on peut être amené à consacrer deux fois plus d'argent et de temps, si ce n'est pas davantage, à une phase ultérieure. Les temps d'arrêt peuvent s'en trouver modifiés et cela peut aussi entraîner l'irradiation de personnes.

Y. SOUSSELIER, France

Il reste encore la question concernant le choix du bitume et celle de savoir s'il est nécessaire de faire des recherches sur les différentes qualités du bitume en fonction, en particulier, de la nature chimique du pétrole brut dont il est extrait. Y a-t-il intérêt à agir, éventuellement dans le cadre d'une coopération internationale, car dans chacun de nos pays nous ne disposons que de sources limitées de pétrole brut alors que nous savons que certains pétroles bruts ont des caractéristiques extrêmement variées ?

G. LEFILLATRE, France

Je pense que dans le domaine du choix du bitume, la difficulté essentielle est justement la très grande variété des origines du pétrole brut. Dans le contexte actuel de crise de l'énergie, les sociétés pétrolières ne peuvent plus, comme il y a une dizaine d'années, garantir une origine bien définie pour leur pétrole. On peut donc avoir des variations très grandes dans les propriétés physico-chimiques des bitumes fournis. Cela pose un problème surtout pour les déchets de plus haute activité, c'est-à-dire la catégorie dont parlait M. Malášek, à savoir au-dessus de 1 Ci/l. Je crois qu'il y a des études à faire pour cette catégorie; on utilise sur les routes des matériaux qui n'ont plus grand chose à voir avec le bitume, mais qui sont des dérivés sulphurés qui ont des propriétés de résistance au vieillissement supérieures à celles du bitume classique. Ceci pourrait faire l'objet d'études pour la fabrication d'un type spécial de bitume, qui pourrait être éventuellement un bitume "vulcanisé", d'études de chimie organique, en collaboration avec les pétroliers, pour vérifier si un tel type de bitume serait nettement supérieur au point de vue résistance aux radiations à ceux que l'on utilise actuellement. A ce moment-là, on pourrait avoir une garantie de fabrication puisqu'on aurait un "bitume atomique" en quelque sorte.

Personnellement, j'ai toujours été un peu surpris de constater que les firmes pétrolières ne savaient pas grand chose sur leur bitume, si ce n'est que c'est un sous-produit pétrolier, et elles n'ont jamais fait d'études théoriques approfondies sur les bitumes, à l'inverse par exemple des polyéthylènes et des produits chimiquement bien définis. Donc, cela suppose des investissements assez lourds puisqu'il faut faire des études théoriques. Jusqu'à présent, à ma connaissance, peut-être sauf en Union soviétique, peu de choses ont été faites dans ce domaine dans le monde, et il y a là matière à études.

E. MALASEK, Tchécoslovaquie

A l'heure actuelle, nous nous inspirons au premier chef, dans l'utilisation du bitume, de considérations d'ordre économique.

Nous nous efforçons d'employer le bitume qui peut normalement être obtenu dans le pays et nous entreprenons alors de vérifier les propriétés de ces produits. D'après les comparaisons effectuées, l'élément le plus décisif n'est pas la source de bitume elle-même, en d'autres termes la source de pétrole, mais la méthode de traitement du pétrole. Compte tenu des conditions prévalant dans notre pays, lorsqu'on a changé de source, les propriétés du bitume ont été relativement peu modifiées et il a été possible d'utiliser le produit de la même installation pour le bitumage des déchets sans rencontrer de difficultés particulières. Bien entendu, il est à nouveau nécessaire de dissocier les déchets de faible activité des déchets de haute activité. Ce qui convient parfaitement à des activités se situant au niveau de 1/10 Ci/kg peut ne pas convenir à des activités plus élevées De ce point de vue, nous jugeons nécessaire de vérifier les propriétés de n'importe quel nouveau type de bitume susceptible d'être produit à l'avenir et de le ranger dans une catégorie qui permette de le comparer aux matières déjà utilisées. Je ne pense pas qu'à l'heure actuelle, dans notre pays, il soit raisonnable de fabriquer certains produits bitumeux uniquement destinés à rendre les déchets insolubles car, dans ce cas nouveau, la question d'économie entre en ligne de compte. Nous pouvons obtenir notre bitume en partie sous forme de produit résiduel et à un prix relativement bas. Si nous demandions à notre industrie de mettre au point une matière spéciale destinée à l'incorporation des déchets, son prix serait au moins cent fois plus élevé. Il vaut mieux utiliser les matières que nous avons et ne pas demander cela à l'industrie.

W. HILD, R.F. d'Allemagne

En ce qui concerne les qualités de bitume sélectionnées pour des opérations d'incorporation particulières, nous avons tous mis au point des procédés d'incorporation à partir de recherches qui avaient pour objet de déterminer le type de bitume convenant à notre cas particulier. On observe à nouveau, en l'occurrence, une interaction entre les caractéristiques des produits et le stockage définitif, par exemple. Nous avons tous effectué des essais dans ce domaine et acquis une très bonne expérience au sujet de ces interactions. Grâce à ces recherches, nous disposons, dans divers pays, de plusieurs installations techniques en service depuis un certain temps qui utilisent différentes qualités de bitume. On a désormais acquis suffisamment d'expérience et, à mon avis, nous sommes pour l'instant en assez bonne position : nous possédons de nombreuses connaissances sur les qualités de bitume à sélectionner et il ne me paraît pas nécessaire, dans l'immédiat, de chercher à trouver de meilleurs produits à base de bitume. Je pense que les données disponibles suffisent largement pour le moment.

H. ESCHRICH, Eurochemic

Le fait que les entreprises pétrolières ne consacrent pas un effort particulier à notre problème tend à prouver que nous ne sommes pas un gros consommateur !

Y. SOUSSELIER, France

Je crois en effet que votre remarque est très judicieuse ! Avant de clore notre session, est-il nécessaire d'ajouter mes conclusions à celles que je vous proposais en ouvrant cette session ? Je crois que ce Séminaire nous a réellement permis de montrer que l'enrobage dans le bitume est un procédé qui fonctionne et qui donne de bons résultats. Nous avons vu bien sûr qu'il reste encore un certain nombre d'incertitudes dont aucune n'est cependant susceptible de mettre en cause le procédé, et je crois que de ce côté-là, nous pouvons continuer à aller franchement de l'avant.

Mesdames, Messieurs, je déclare clos le Séminaire sur l'incorporation dans le bitume.

List of Participants

Liste des Participants

AUSTRIA - AUTRICHE

JAKUSCH, H., Vereinigte Edelstahlwerke AG., Forschungszentrum, 2444 Seibersdorf

KNOTIK, K., Dr., Institut für Chemie, Österreichische Studiengesellschaft für Atomenergie GmbH, Forschungszentrum, 2444 Seibersdorf

BELGIUM - BELGIQUE

CANTILLON, G.E., Institut d'Hygiène et d'Epidémiologie, Section Radioactivité, 14 rue Juliette Wytsman, 1050 Bruxelles

DEJONGHE, P., Dr., Directeur général adjoint, SCK/CEN, Boeretang 200, 2400 Mol-Donk

SHANK, E., Comprimo België N.V., Noorderlaan 139, 2030 Anvers

STORRER, J., Directeur adjoint, BelgoNucléaire S.A., 25 rue du Champ de Mars, 1050 Bruxelles

VAN DE VOORDE, N., Chef de Département principal des installations de traitement de résidus radioactifs, BelgoNucléaire, SCK/CEN, Boeretang 200, 2400 Mol-Donk

CANADA

HEATHCOCK, R.E., Ontario Hydro, Design & Development Division H14, Toronto, Ontario M56 IX6

WILLIAMSON, A.S., Ontario Hydro, Research Division, 800 Kipling Avenue, Toronto, Ontario M8Z 5S4

DENMARK - DANEMARK

BATSBERG PEDERSEN, Research Establishment Risø, 4000 Roskilde

BRODERSEN, K., Research Establishment Risø, 4000 Roskilde

FINLAND - FINLANDE

AITTOLA, J-P., Technical Research Centre of Finland, Reactor Laboratory, 02150 Espoo

HANELIUS, A., TVO Power Company, Kutojantie 8, 02180 Espoo 18

SÖDERMAN, J.K., Oy Rosenlew AB, Engineering Works, P.O. Box 51, 28101 Pori 10

FRANCE

FERNANDEZ, N., Chef de la Section de traitement des effluents et des déchets, Commissariat à l'Energie Atomique, Centre d'Etudes Nucléaires de Marcoule, B.P. n° 106, 30200 Bagnols-sur-Cèze

LE BLAYE, G., Saint Gobain Techniques Nouvelles, 23 boulevard Georges Clémenceau, 92400 Courbevoie

LECONNETABLE, J., Responsable de la station d'enrobage par le bitume, Service de protection contre les rayonnements, Section IDS, Commissariat à l'Energie Atomique, Centre d'Etudes Nucléaires de Saclay, B.P. n° 2, 91190 Gif-sur-Yvette

LEFILLATRE, G., Service de Chimie Appliquée, Commissariat à l'Energie Atomique, Centre d'Etudes Nucléaires de Cadarache, B.P. n° 1, 13115 Saint-Paul-lez-Durance

SOUSSELIER, Y., Adjoint au Directeur du Plutonium, Direction des Productions, Commissariat à l'Energie Atomique, B.P. n° 74, 92320 Chatillon-sous-Bagneux

FEDERAL REPUBLIC OF GERMANY - REPUBLIQUE FEDERALE D'ALLEMAGNE

BREGULLA, E., Niklear-Chemie und Metallurgie GmbH., Postfach 869, 6450 Wolfgang bei Hanau

ENGELHARDT, G., Gesellschaft für Kernforschung mbH, Projektbereich WAK, Postfach 3640, 7500 Karlsruhe

HILD, W., Dr., Nuclear Research Center, Waste Management Research Department, Gesellschaft für Kernforschung mbH., Weberstrasse 5, 7500 Karlsruhe

KLÜGER, V., Gesellschaft für Kernforschung mbH., ABRA, Weberstrasse 5, 7500 Karlsruhe

SMAILOS, E., Gesellschaft für Kernforschung mbH., ABRA, Weberstrasse 5, 7500 Karlsruhe

UERPMANN, E.P., Institut für Tieflagerung, Berlinerstrasse 2, 3392 Clausthal-Zellerfeld

ITALY - ITALIE

ANTONUCCI, L., Dr., Ente Nazionale per l'Energia Elettrica, Direzionne Produzzione e Trasmissione, Settore Nucleare, Viale Regina Margherita 137, 00198 Rome

CESTARO, G., Comitato Nazionale per l'Energia Nucleare, Impianto Eurex, 13040 Saluggia (Vercelli)

DWORSCHAK, H., Comitato Nazionale per l'Energia Nucleare, Direttore Impianto Eurex, 13040 Saluggia (Vercelli)

RISOLUTI, P., Dr., AGTP Nucleare, Corso di Porta Romana 68, 20122 Milano

ZIFFERERO, M., Prof., Comitato Nazionale per l'Energia Nucleare, Viale Regina Margherita 125, 00198 Rome

JAPAN - JAPON

MATSUURA, H., NAIG Co., Ukishimacho -4-1, Kawasakiku, Kawasaki, Kanagawa

SEGAWA, T., Dr., Tokai Works PNC, 3371 Muramatsu, Tokai-Mura, Naka-Gun, Ibaraki-Ken

THE NETHERLANDS - PAYS-BAS

HOEFNAGELS, J.H.C., N.V. KEMA, Utrechtseweg 310, Arnhem

SMEETS, L., Waste Treatment, Reactor Centrum Nederland, Westerduinweg 3, Petten (N.H.)

VAN DER PLAS, T., N.V. KEMA, Utrechtseweg 310, Arnhem

NORWAY - NORVEGE

TALLBERG, K.A., Institutt for Atomenergi, P.O. Box 40, 2007 Kjeller

SPAIN - ESPAGNE

ORTEGA ABELLAN, J., Dr., Division des combustibles irradiés, Junta de Energia Nuclear, Avenida Complutense 28, Madrid 3

SWEDEN - SUEDE

CHRISTENSEN, H., Department KVC, AB ASEA-ATOM, P.O. Box 53, 721 04 Västerås

EDWALL, B., South Swedish Power Company, Fack, 200 70 Malmø 5

GUSTAFSSON, B.J.F., Swedish Nuclear Fuel Supply Co., Brahegatan 47, Fack, 102 40 Stockholm 5

LINDEROTH, G., Swedish State Power Board, Konstruktions sektionen för Värmeteknik, 162 87 Vällingby

TOLLBÄCK, H., AB Atomenergi, Fack 611 01 Nyköping 1

SWITZERLAND - SUISSE

AEPPLI, J., Abt. für die Sicherheit der Kernanlagen, Sektion Personen- und Umgebungsschutz, 5303 Würenlingen

BURRI, H.R., Electrowatt Engineering, Bellerivestrasse 36, 8022 Zürich

KNUTTI, P.W., Motor Colombus AG, Parkstrasse 23, 5401 Baden

UNITED KINGDOM - ROYAUME-UNI

CLARKE, J.H., UK Atomic Energy Authority, Atomic Energy Research Establishment, Process Technology Division, Building 175, Harwell, Didcot, Oxfordshire OX11 ORA

HELSBY, G.H., Nuclear Installations Inspectorate, Branch 3, Silkhouse Court, Tithebarn Street, Liverpool L2 2LZ

JEAPES, A.P., Dr., British Nuclear Fuels Limited, Research and Development Department, Windscale Works, Seascale, Cumbria CA20 IPG

UNITED STATES - ETATS-UNIS

STEWART, J.E., Manager, Werner & Pfleiderer Corp., 160 Hopper Avenue, Waldwick, N.J. 07463

COMMISSION OF THE EUROPEAN COMMUNITIES - COMMISSION DES COMMUNAUTES EUROPEENNES

GRITTI, R., CCR Euratom, Département Chimie, 20122 Ispra (Italy)

SENTOLL, J., CCR Euratom, Département Chimie, 20122 Ispra (Italy)

SIMON, R., Commission of the European Communities, 200 rue de la Loi, 1040 Brussels (Belgium)

INTERNATIONAL ATOMIC ENERGY AGENCY - AGENCE INTERNATIONALE DE L'ENERGIE ATOMIQUE

JOURDE, P., Section gestion des déchets, Division de la sûreté nucléaire et de la protection de l'environnement, AIEA, Kärntnerring 11, 1011 Vienna

MALÁŠEK, E., Dr., Czechoslovac Atomic Energy Commission, Slezska 9, Prague 2 (Czechoslovakia)

EUROCHEMIC COMPANY - SOCIETE EUROCHEMIC

DETILLEUX, E., Managing Director, Boeretang 200, 2400 Mol-Donk

ESCHRICH, H., Deputy Manager

BALSEYRO CASTRO, M., Development Department

DEMONIE, M., Plant Operation Department

FRANKIGNOUL, N., General Services Department

GARCIA GALAN, R., Development Department

HUMBLET, L., Development Department

OSIPENCO, A., Health & Safety Department
PIVATO, J., Development Department
REYNDERS, R., Plant Operation Department
SCARABELLI, R., Stagiaire, Plant Operation Department
STERNER, H., Stagiaire, Plant Operation Department
VAN GEEL, J., Development Department

OECD NUCLEAR ENERGY AGENCY -
AGENCE DE L'OCDE POUR L'ENERGIE NUCLEAIRE

SAELAND, E., Director General, 38 boulevard Suchet, 75016 Paris

SECRETARIAT OF THE SEMINAR - SECRETARIAT DU SEMINAIRE

DRENT, W., Société Eurochemic, Boeretang 200, 2400 Mol-Donk

OLIVIER, J-P., Agence de l'OCDE pour l'Energie Nucléaire,
38 boulevard Suchet, 75016 Paris

JACOBS, G., Mme, Société Eurochemic, Boeretang 200, 2400 Mol-Donk

KOUSNETZOFF, C., Mme, Agence de l'OCDE pour l'Energie Nucléaire,
38 boulevard Suchet, 75016 Paris

Some other publications of NEA

ACTIVITY REPORTS

Activity Reports of the OECD Nuclear Energy Agency (NEA)	Second Activity Report (1973) 71 pages (crown 4to) Third Activity Report (1974) 75 pages (crown 4to) Fourth Activity Report (1975) 77 pages (crown 4to)

Free on request

Annual Reports of the OECD High Temperature Reactor Project (DRAGON)	Fourteenth Report (1972-1973) 112 pages (crown 4to) Fifteenth Report (1973-1974) 85 pages (crown 4to) Sixteenth Report (1974-1975) 99 pages (crown 4to)

Free on request

Annual Reports of the OECD Halden Reactor Project	Thirteenth Report (1972) 178 pages (crown 4to) Fourteenth Report (1973) 105 pages (crown 4to) Fifteenth Report (1974) 103 pages (crown 4to)

Free on request

SCIENTIFIC AND TECHNICAL CONFERENCE PROCEEDINGS

Radiation Dose Measurements (Their purpose, interpretation and required accuracy in radiological protection)	Proceedings of the Stockholm Symposium, June 1967 597 pages (crown 4to) 64s., $ 11, F 44, FS 44, DM 36.50
Application of On-Line Computers to Nuclear Reactors	Proceedings of the Sandefjord Seminar, September 1968 900 pages (crown 4to) £ 7.5s., $ 20, F 85, FS 78, DM 70
Third Party Liability and Insurance in the Field of Maritime Carriage of Nuclear Substances	Proceedings of the Monaco Symposiu[m] October 1968 529 pages (crown 8vo) £ 2.12s., $ 7.50, F 34, FS 28.50, DM 22.50
The Physics Problems of Reactor Shielding	Proceedings of the Specialist Meeting, Paris, December 1970 175 pages (crown 4to) £ 1.75, $ 5, F 23, FS 20, DM 15.60
Magnetohydrodynamic Electrical Power Generation	Proceedings of the Fifth International Conference, Münich, April 1971 499 pages (crown 4to) £ 4.88, $ 14, F 65, FS 50, DM 43
Marine Radioecology	Proceedings of the Hamburg Seminar September 1971 213 pages (crown 8vo) £ 1.50, $ 4.50, F 20, FS 15,60, DM 13.60
Disposal of Radioactive Waste	Proceedings of the Information Meeting, Paris, 12th-14th April 19[illegible] 290 pages (crown 8vo) £ 2.60, $ 7.75, F 32, FS 25, DM 20
Power from Radioisotopes	Proceedings of the Second International Symposium, Madrid, 29th May-1st June 1972 986 pages (crown 4to) £ 9, $ 24, F 110, FS 83.50, DM 68.8[illegible]
The Management of Radioactive Wastes from Fuel Reprocessing	Proceedings of the Paris Symposium, 27th November-1st December 1972 1265 pages (crown 8vo) £ 12, $ 34, F 140, FS 107, DM 88
The Monitoring of Radioactive Effluents	Proceedings of the Karlsruhe Seminar, 14th-17th May 1974 452 pages (crown 8vo) £ 4.40, $ 11, F 44

Management of Plutonium-Contaminated Solid Wastes	Proceedings of the Marcoule Seminar, 14th-16th October 1974 248 pages (crown 8vo) £ 3.80, $ 9.50, F 38
Bituminization of Low and Medium Level Radioactive Wastes	Proceedings of the Antwerp Seminar 18th-19th May 1976 in preparation

SCIENTIFIC AND TECHNICAL REPORTS

Radiation Protection Norms	Revised Edition 1968 Free on request
Radioactive Waste Disposal Operation into the Atlantic 1967	September 1968 74 pages (crown 8vo) 12s., $ 1.80, F 8, FS 7, DM 5.80
Power Reactor Characteristics	September 1966 83 pages (crown 4to) 15s., $ 2.50, F 10, FS 10, DM 8.30
Uranium Resources (Revised Estimates)	December 1967 27 pages (crown 4to) Free on request
Prospects for Nuclear Energy in Western Europe : Illustrative Power Reactor Programmes	May 1968 47 pages (crown 4to) 17s.6d., $ 2.50, F 10, FS 10, DM 8.30
Uranium - Production and Short Term Demand	January 1969 29 pages (crown 4to) 7s., $ 1, F 4, FS 4, DM 3.30
Uranium - Resources, Production and Demand	September 1970 54 pages (crown 4to) £ 1, $ 3, F 13, FS 11.50, DM 9.10
Uranium - Resources, Production and Demand	August 1973 140 pages (crown 4to) £ 1.76, $ 5, F 20, FS 15.60, DM 12.50
Uranium - Resources, Production and Demand, including other Nuclear Fuel Cycle Data	December 1975 78 pages (crown 4to) £ 3.10, $ 7, F 28
Water Cooled Reactor Safety	May 1970 179 pages (crown 4to) £ 1.52, $ 4.50, F 20, FS 17.50, DM 13.60

Glossary of Terms and Symbols in Thermionic Conversion	1971 112 pages (crown 4to) £ 1.75, $ 5, F 23, FS 20, DM 15.60
Radioactive Waste Management Practices in Western Europe	1972 126 pages (crown 8vo) £ 1.15, $ 3.25, F 15, FS 11.70, DM 10.50
Radioactive Waste Management Practices in Japan	1974 45 pages (crown 8vo) Free on request
Basic Approach for Safety Analysis and Control of Products Containing Radionuclides and Available to the General Public	June 1970 31 pages (crown 8vo) 11s., $ 1.50, F 7, FS 6, DM 4.90
Radiation Protection Standards for Gaseous Tritium Light Devices	1973 23 pages (crown 8vo) Free on request
Radiation Protection Considerations on the Design and Operation of Particle Accelerators	1974 80 pages (crown 8vo) Free on request
Interim Radiation Protection Standards for the Design, Construction, Testing and Control of Radioisotopic Cardiac Pacemakers	1974 54 pages (crown 8vo) £ 1, $ 2.50, F 10
Guidelines for Sea Disposal Packages of Radioactive Waste	November 1974 32 pages (crown 8vo) Free on request
Estimated Population Exposure from Nuclear Power Production and Other Radiation Sources	January 1976 48 pages (crown 8vo) £ 1.60, $ 3.50, F 14

LEGAL PUBLICATIONS

Convention on Third Party Liability in the Field of Nuclear Energy	July 1960, incorporating provisions of Additional Protocol of January 1964 73 pages (crown 4to) Free on request
Nuclear Legislation, Analytical Study : "Nuclear Third Party Liability"	1967 78 pages (crown 8vo) 14s., $ 2.30, F 9, FS 9, DM 7.50 (it is planned to publish a revised version of this study)
Nuclear Legislation, Analytical Study : "Organisation and General Regime Governing Nuclear Activities"	1969 230 pages (crown 8vo) £ 2, $ 6, F 24, FS 24, DM 20

Nuclear Legislation, Analytical Study : "Regulations Governing Nuclear Installation and Radiation Protection"	1972 492 pages (crown 8vo) £ 3.70, $ 11, F 45, FS 34.60, DM 29.80
Nuclear Law Bulletin	Annual Subscription Two issues and supplements £ 2.80, $ 6.25, F 25

OECD SALES AGENTS
DEPOSITAIRES DES PUBLICATIONS DE L'OCDE

ARGENTINA - ARGENTINE
Carlos Hirsch S.R.L.,
Florida 165, BUENOS-AIRES.
☎ 33-1787-2391 Y 30-7122

AUSTRALIA - AUSTRALIE
International B.C.N. Library Suppliers Pty Ltd.,
161 Sturt St., South MELBOURNE, Vic. 3205.
☎ 69.7601
658 Pittwater Road, BROOKVALE NSW 2100.
☎ 938 2267

AUSTRIA - AUTRICHE
Gerold and Co., Graben 31, WIEN 1. ☎ 52.22.35

BELGIUM - BELGIQUE
Librairie des Sciences
Coudenberg 76-78, B 1000 BRUXELLES 1.
☎ 512-05-60

BRAZIL - BRESIL
Mestre Jou S.A., Rua Guaipá 518,
Caixa Postal 24090, 05089 SAO PAULO 10.
☎ 216-1920
Rua Senador Dantas 19 s/205-6, RIO DE JANEIRO GB. ☎ 232-07. 32

CANADA
Publishing Centre/Centre d'édition
Supply and Services Canada/Approvisionnement et Services Canada
270 Albert Street, OTTAWA K1A 0S9, Ontario
☎ (613)992-9738

DENMARK - DANEMARK
Munksgaards Boghandel
Nørregade 6, 1165 KØBENHAVN K.
☎ (01) 12 69 70

FINLAND - FINLANDE
Akateeminen Kirjakauppa
Keskuskatu 1, 00100 HELSINKI 10. ☎ 625.901

FRANCE
Bureau des Publications de l'OCDE
2 rue André-Pascal, 75775 PARIS CEDEX 16.
☎ 524.81.67
Principaux correspondants :
13602 AIX-EN-PROVENCE : Librairie de l'Université. ☎ 26.18.08
38000 GRENOBLE : B. Arthaud. ☎ 87.25.11
31000 TOULOUSE : Privat. ☎ 21.09.26

GERMANY - ALLEMAGNE
Verlag Weltarchiv G.m.b.H.
D 2000 HAMBURG 36, Neuer Jungfernstieg 21
☎ 040-35-62-500

GREECE - GRECE
Librairie Kauffmann, 28 rue du Stade,
ATHENES 132. ☎ 322.21.60

HONG-KONG
Government Information Services,
Sales of Publications Office,
1A Garden Road,
☎ H-252281-4

ICELAND - ISLANDE
Snaebjörn Jónsson and Co., h.f.,
Hafnarstraeti 4 and 9, P.O.B. 1131,
REYKJAVIK. ☎ 13133/14281/11936

INDIA - INDE
Oxford Book and Stationery Co. :
NEW DELHI, Scindia House. ☎ 47388
CALCUTTA, 17 Park Street. ☎ 24083

IRELAND - IRLANDE
Eason and Son, 40 Lower O'Connell Street,
P.O.B. 42, DUBLIN 1. ☎ 74 39 35

ISRAEL
Emanuel Brown :
35 Allenby Road, TEL AVIV. ☎ 51049/54082
also at :
9, Shlomzion Hamalka Street, JERUSALEM.
☎ 234807
48 Nahlath Benjamin Street, TEL AVIV.
☎ 53276

ITALY - ITALIE
Libreria Commissionaria Sansoni :
Via Lamarmora 45, 50121 FIRENZE. ☎ 579751
Via Bartolini 29, 20155 MILANO. ☎ 365083
Sous-dépositaires :
Editrice e Libreria Herder,
Piazza Montecitorio 120, 00186 ROMA.
☎ 674628
Libreria Hoepli, Via Hoepli 5, 20121 MILANO.
☎ 865446
Libreria Lattes, Via Garibaldi 3, 10122 TORINO.
☎ 519274
La diffusione delle edizioni OCDE è inoltre assicurata dalle migliori librerie nelle città più importanti.

JAPAN - JAPON
OECD Publications Centre,
Akasaka Park Building,
2-3-4 Akasaka,
Minato-ku
TOKYO 107. ☎ 586-2016
Maruzen Company Ltd.,
6 Tori-Nichome Nihonbashi, TOKYO 103,
P.O.B. 5050, Tokyo International 100-31.
☎ 272-7211

LEBANON - LIBAN
Documenta Scientifica/Redico
Edison Building, Bliss Street,
P.O.Box 5641, BEIRUT. ☎ 354429 - 344425

THE NETHERLANDS - PAYS-BAS
W.P. Van Stockum
Buitenhof 36, DEN HAAG. ☎ 070-65.68.08

NEW ZEALAND - NOUVELLE-ZELANDE
The Publications Manager,
Government Printing Office,
WELLINGTON : Mulgrave Street (Private Bag),
World Trade Centre, Cubacade, Cuba Street,
Rutherford House, Lambton Quay ☎ 737-320
AUCKLAND : Rutland Street (P.O.Box 5344)
☎ 32.919
CHRISTCHURCH : 130 Oxford Tce, (Private Bag)
☎ 50.331
HAMILTON : Barton Street (P.O.Box 857)
☎ 80.103
DUNEDIN : T & G Building, Princes Street
(P.O.Box 1104), ☎ 78.294

NORWAY - NORVEGE
Johan Grundt Tanums Bokhandel,
Karl Johansgate 41/43, OSLO 1. ☎ 02-332980

PAKISTAN
Mirza Book Agency, 65 Shahrah Quaid-E-Azam,
LAHORE 3. ☎ 66839

PHILIPPINES
R.M. Garcia Publishing House,
903 Quezon Blvd. Ext., QUEZON CITY,
P.O. Box 1860 - MANILA. ☎ 99.98.47

PORTUGAL
Livraria Portugal,
Rua do Carmo 70-74. LISBOA 2. ☎ 360582/3

SPAIN - ESPAGNE
Libreria Mundi Prensa
Castelló 37, MADRID-1. ☎ 275.46.55
Libreria Bastinos
Pelayo, 52, BARCELONA 1. ☎ 222.06.00

SWEDEN - SUEDE
Fritzes Kungl. Hovbokhandel,
Fredsgatan 2, 11152 STOCKHOLM 16.
☎ 08/23 89 00

SWITZERLAND - SUISSE
Librairie Payot, 6 rue Grenus, 1211 GENEVE 11.
☎ 022-31.89.50

TAIWAN
Books and Scientific Supplies Services, Ltd.
P.O.B. 83, TAIPEI.

TURKEY - TURQUIE
Librairie Hachette,
469 Istiklal Caddesi,
Beyoglu, ISTANBUL, ☎ 44.94.70
et 14 E Ziya Gökalp Caddesi
ANKARA. ☎ 12.10.80

UNITED KINGDOM - ROYAUME-UNI
H.M. Stationery Office, P.O.B. 569, LONDON
SE1 9 NH, ☎ 01-928-6977, Ext. 410
or
49 High Holborn
LONDON WC1V 6HB (personal callers)
Branches at: EDINBURGH, BIRMINGHAM, BRISTOL, MANCHESTER, CARDIFF, BELFAST.

UNITED STATES OF AMERICA
OECD Publications Center, Suite 1207,
1750 Pennsylvania Ave, N.W.
WASHINGTON, D.C. 20006. ☎ (202)298-8755

VENEZUELA
Libreria del Este, Avda. F. Miranda 52,
Edificio Galipán, Aptdo. 60 337, CARACAS 106.
☎ 32 23 01/33 26 04/33 24 73

YUGOSLAVIA - YOUGOSLAVIE
Jugoslovenska Knjiga, Terazije 27, P.O.B. 36,
BEOGRAD. ☎ 621-992

Les commandes provenant de pays où l'OCDE n'a pas encore désigné de dépositaire peuvent etre adressées à :
OCDE, Bureau des Publications, 2 rue André-Pascal, 75775 Paris CEDEX 16
Orders and inquiries from countries where sales agents have not yet been appointed may be sent to
OECD, Publications Office, 2 rue André-Pascal, 75775 Paris CEDEX 16

PUBLICATIONS DE L'OCDE, 2, rue André-Pascal, 75775 Paris Cedex 16 - No. 37 541 1976

IMPRIMÉ EN FRANCE